Springer Series in

SOLID-STATE SCIENCES 162

Springer Series in

SOLID-STATE SCIENCES

Series Editors:
M. Cardona P. Fulde K. von Klitzing R. Merlin H.-J. Queisser H. Störmer

The Springer Series in Solid-State Sciences consists of fundamental scientific books prepared by leading researchers in the field. They strive to communicate, in a systematic and comprehensive way, the basic principles as well as new developments in theoretical and experimental solid-state physics.

Please view available titles in *Springer Series in Solid-State Sciences* on series homepage http://www.springer.com/series/682

Karl W. Böer

Electro-Optical Effects to Visualize Field and Current Distributions in Semiconductors

With 74 Figures

Springer

Professor Dr. Karl W. Böer
University of Delaware
Dept. Physics & Astronomy
Newark DE 19711
USA
E-mail: solpax@aol.com

Series Editors:

Additional material to this book can be downloaded from http://extras.springer.com

Springer Series in Solid-State Sciences ISSN 0171-1873
ISBN 978-3-642-03439-8 ISBN 978-3-642-03440-4 (eBook)
DOI 10.1007/978-3-642-03440-4
Springer Heidelberg Dordrecht London New York

Library of Congress Control Number: 2009939437

Cover design: SPi Publisher Services

Printed on acid-free paper

Springer is part of Springer Science+Business Media (www.springer.com)

Preface

This monograph of *Electro-Optical Effects to Visualize Field- and Current-Distributions in Semiconductors* consists of five parts, four of which are based on the research of cadmium sulfide, where a large number of contributions were made between 1958 and the late 1960s to directly observe field and current distributions and interpret their results. The visualization of field distributions was accomplished by using the Franz Keldysh effect, and the visualization of current inhomogeneities uses the shift of the optical absorption edge by Joule's heating. The fifth part deals with a review of the explosively developing field of N- and S-shaped current voltage characteristics causing inhomogeneities and instabilities in field and current distributions. This part of the book was composed by Eckehard Schöll of the Technical University in Berlin.

A major emphasis is given to the first part of the book in which stationary high-field domains are described. These domains can be used as an essential tool to determine unambiguously certain semiconductor properties, such as the electron density and its mobility as the function of the actual electric field. It is also helpful to determine changes of the work function and electron affinities between different materials, such as for electrodes and heterojunctions. Finally, it gives direct information about certain doping and their spacial profile.

In the second part of the book, we describe kinetic effects of moving high-field domains that can be followed directly by observation with the help of the Franz Keldysh effect. Because such kinetic description is limited in any book by presenting a few snapshots and its theory, we have augmented this part by an appendix part, which describes a film that is now available from the Web site of Springer Verlag. It is heavily referenced in all relevant sections of the book and gives a most instructive presentation of the actual kinetics of the effects in real time frames.

The kinetic effects of certain high-field domain effects have significant technical interest for high frequency current generation, for example, via the Gunn effect, or are of interest to analyze in more detail some of the reasons for excessive electronic noise at higher operating voltages.

More recently, cadmium sulfide has gained new interest as an extended electrode to many thin-film solar cells, such as the CdS/CdTe or $CdS/CuInSe_2$ cell. These cells show substantially improved conversion efficiencies when CdS is used as a thin interlayer to any other conventional electrode. Its ability to form high-field domains in the range of negative differential conductivity and to change from n- to p-type in such domains close to a p-type photoconductor may be the key for explaining its usefulness. This is one more reason to pay specific attention to the explanation of such effects here.

The third part of the book deals with current channels that develop at higher current densities in the S-shaped negative resistivity range. These current channels can be seen by a shift of the optical absorption edge caused by Joule's heating of the channel. The channels often move between the electrodes. This is described in several sections and more dramatically presented in the second part of the film, that is, on the Web site of the Springer Verlag. Again this is heavily referenced within the book and also include a large variety of pre-breakdown phenomena, again shown in the example of cadmium sulfide.

The dramatic growth of the field in the following decades demanded a careful review of the many evolving results and effects that were comprehensively presented by E. Schöll in the fifth and sixth part of the book. In these parts it was felt that also a more general analytical discussion of the effects caused by negative differential conductivities is called for.

With all this, we hope to have given a unique opportunity to the reader to become familiar with the basics of field and current inhomogeneities and instabilities, and to use some of them as useful tools and means to create or improve semiconductor devices.

I acknowledge the dedicated help I received from Ms. Anita Schwartz of the Information Technology Department of the University of Delaware to assist me in composing the text of this book in electronic form.

My special gratitude goes to Renate, my wife, who expected me to be truly retired and spend more time relaxing with her, while I was most of the days in my office, trying to find the proper way to explain in writing to future students and colleagues the intricacy of field and current instabilities in semiconductors and their relevance to current advances in the technology of solar cells.

Newark, *Karl W. Böer*
November 2009

Contents

1

Introduction

1.1 A Brief History

When the section of the Dielectric Breakdown of the German Academy of Science in Berlin was created, we were to investigate the electrical behavior of single crystals, mainly cadmium sulfide at high-electric fields in the pre-breakdown range. The only electrical tool we had at that time was the analysis of the current–voltage characteristics that started to show significant nonlinear deviations from the behavior at low fields. However, this characteristic is an insufficient tool to make important conclusions of the behavior of the field dependence of the carrier motion within the crystal. The measured current essentially integrates over all current channels within a crystal; and the field when dividing the applied voltage by the distance between the electrodes again is an averaged one that says little about the microscopic processes that create the nonlinearities of the current–voltage characteristics that we observed.

To obtain the first information about the field and current distributions, we had to resort to a tedious point for point probing with a thin metal wire along the surface, and it was possible to an acceptable degree of accuracy only on thin crystal platelets that left sufficient space between the electrodes for this probing (see Fig. 1.1). Usually, after hours of such point for point probing, a profile of equi-potential lines and their orthogonal trajectories showing the current distribution could be assembled.

The first few weeks of the Spring of 1958 of my research group were filled with the investigation of the current through CdS at higher applied voltages and the concurrent analysis of the field and current profiles using such probing methods. It showed a steeper potential drop in front of the cathode, known as a Schottky barrier and a rather uneventful linear potential distribution in the center except for another (less) steep increase before the anode.

More extensive probing of the entire potential profile, though rather time consuming, did not disclose more interesting details except for some inhomogeneities caused by crystal defects, but could not explain some of the changes we observed. The experimental method was too slow to explain some kinetic

Fig. 1.1. Typical crystal platelet with evaporated electrodes offset and positioned on opposite sides to avoid surface glide discharges at higher applied voltages

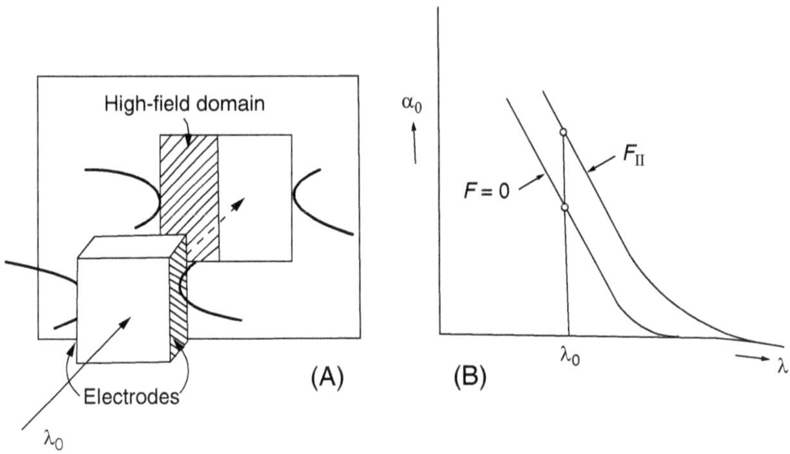

Fig. 1.2. Schematics of the experimental setup to observe field inhomogeneities using electro-optical effects

effects and was also designed only to measure the potential distribution along the surface of the crystal.

In search of a probe that could detect the electric field within the crystal, we knew about the Stark effect that would deform the absorption spectrum of a defect center, and the inverse piezo effect that would change the lattice constant and thereby shift the band edge.

We decided to look through the crystal at low temperatures close to the optical absorption edge (Fig. 1.2). Ulrich Kümmel, one of my associates, conducted the first experiment on a thin copper-doped CdS crystal with two small circular electrodes, offset on different sides of the crystal. The available vacuum vessel with windows on each side permitted lowering the temperature to $-70°$C. His first observations were totally unexpected: Not some darkening due to expected fields between the electrodes, but a dark ring was observed *around* the cathode when the applied voltage was high enough.

That ring increased in diameter, separated itself from the cathode in time, grew until it reached at one point the anode (Fig. 1.3), then disappeared and a new ring was formed around the cathode and the process repeated itself with a periodicity of about 10 s. At the same time, the current through the crystal decreased while the ring was seen and increased again to about twice its value when the ring disappeared: the current oscillated.

(a) (b) (c)

Fig. 1.3. Sequence of photographs of the CdS crystal in monochromatic light close to the optical absorption edge, showing a dark ring that increases in diameter in time; pictures are taken 2 s apart. The end of the electrode wires present small circle electrode cathode on upper right, anode on lower left

We found out later that we were fortunate to observe this phenomenon, as the temperature was just in the right range, as we will describe later, and the crystal had the right Cu-doping and the light intensity was sufficiently small to permit an applied voltage of several 100 V before breakdown. We repeated the experiment and convinced ourselves that this was a true volume effect of an *electric field domain* (as we called it). But our explanation was wrong, it was obviously not caused by the much smaller Stark effect and also not by the inverse piezo effect. In the anisotropic CdS with its c-axis within the platelet we should have observed dark sickle-shaped features and no rings.

Fortunately around the same time, a paper of Walter Franz that appeared described a lowering of the band edge in sufficient electric fields by tunneling of electrons slightly into the band gap (Franz 1958). A little later, and independent of Franz, Keldysh described the same effect(Keldysh 1958). We stumbled into it and consequently coined it the Franz–Keldysh effect that helped us to discover the first high-field domain and its related instability (Böer et al. 1958, 1959, 1961). Though sufficiently slow to observe the motion with our eye, we would never have observed it with the electrode probing.

This event caused a reorganization of my research group with major efforts to study more of these field inhomogeneities. Soon thereafter, at higher light intensities and higher currents, current channels occurred that were easily observed by the much larger shift of the optical absorption edge due to Joule's heating. A systematic research activity started, first sporadically observing one or the other phenomenon, and slowly bringing order into the system involving carriers, space charges and the resulting field and current distribution in the range of negative differential conductivities. We will describe this systematically in later chapters.

However, since most of the inhomogeneous field and current distributions are of kinetic nature, we recorded all of these on a film (Böer at al. 1961). We later added to the film explanatory material that described in more detail the involved effects. This film was assembled as an educational film, shown at many conferences and in class rooms. Since it describes most of the kinetic material presented in this book in greater detail, we offer it complimentary to the text on the Springer Web site and will, from time to time refer to it in the following chapters. The first part of the film also shows an instructive example of the probing method described earlier and a typical orthogonal trajectory picture of the observed equi-potential lines and current distributions.

1.2 Negative Resistance Regimes

With sufficiently increased applied voltages the current–voltage characteristics shows significant changes from the low-voltage behavior. Depending on the conductivity, the characteristic may become N-shaped, by first increase, than after reaching a maximum decrease, and at still higher voltages increase again as shown in Fig. 1.4a. When the conductivity of the semiconductor is higher, the characteristic becomes S-shaped as shown in Fig. 1.4b, as a current channel forms before the field is high enough for the N-shaped regime to occur.

In photoconducting CdS a shift from the N- to the S-shaped regime can easily be achieved by increasing the optical excitation that permitted us to separate both and investigate them separately.[1]

These typical characteristics pointed out by Ridley (1963), however, are rarely observed in the above given simple form, as instabilities often occur that

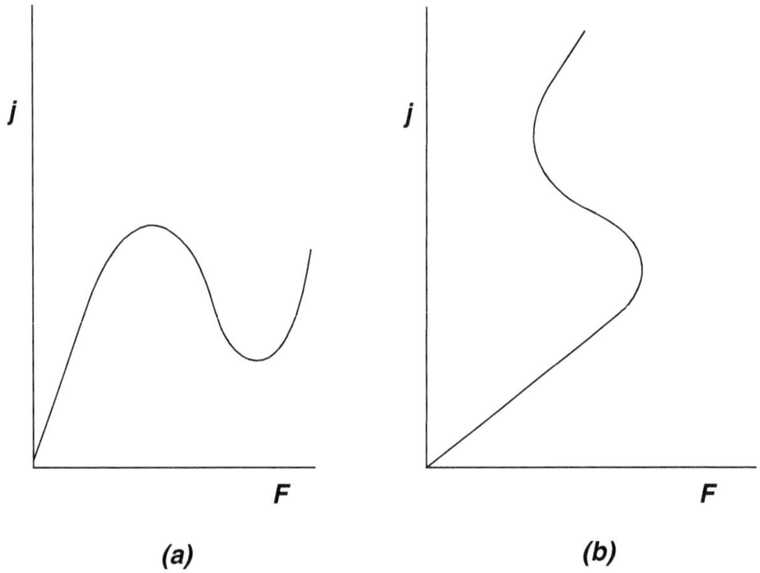

Fig. 1.4. Typical current–voltage characteristics (schematic) in the N-shaped regime shown in (**a**) and in the S-shaped regime shown in (**b**)

[1] As well as the more complicated transition range between these regimes with intermediate optical excitation, that is, in an intermediate conductivity range.

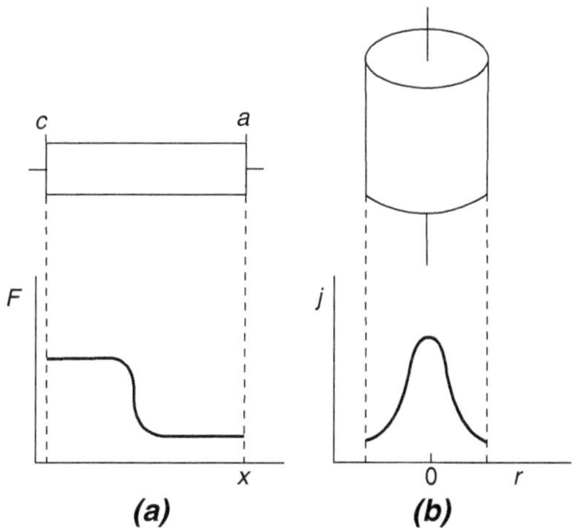

Fig. 1.5. (**a**) Platelet-shaped specimen to investigate high-field domains (**b**) cylinder-shaped specimen to investigate current channels in semiconductors

complicate this picture. For the N-shaped regime, typical high-field domains appear that render the current either saturating rather decreasing or causing current oscillations. For the S-shaped regime to fully develop, one needs to employ a sufficiently large series resistance to avoid a run-away thermal breakdown.

For these reasons it is more important to focus on the conductivity as a function of the applied voltage, or better on the acting field in the N-shaped regime, and on the actual current distribution in the S-shaped regime. For this purpose, we will choose specific geometries of the samples that render the analysis more simple: Platelets for the N-shaped regime that result in high-field domains and a cylinder for the theoretical analysis of the S-shaped regime that result in current channels (Fig. 1.5). We will do the latter analysis in Chap. 6 of this book.

We will first analyze the N-shaped regime in more detail. Here, the conductivity decreases with increasing fields, either by decreasing carrier density (as e.g., in CdS) or by decreasing mobility (as in GaAs).

When the conductivity decreases stronger than linearly with increasing field, typical high-field domains will occur. To understand this behavior a detailed analysis of this is necessary which will be given in Chap. 2.

1.3 Summary and Emphasis

The Franz–Keldysh effect is useful as an electro-optical method to make field inhomogeneities directly visible and observe a new type of field inhomogeneities, the high-field domain, that can move from cathode to anode in CdS, causing current oscillations.

Such domains occur when the conductivity decreases stronger than linearly with increasing electric field. This can occur when the carrier density or the mobility decreases with increasing field. The first effect is observed in CdS, the second one in GaAs.

High-field domains are regions within the semiconductor with essentially constant field that is substantially above the field in the rest of the crystal.

2

High-Field Domains

Summary. In regimes of N-shaped negative differential conductivity, high-field domains are created which can be stationary or move through the device and cause a reduction of current or self-generated current oscillations. The latter, as in the Gunn effect, or in super lattices, may be used as a high-frequency ac generator.

Homogeneously doped semiconductors develop inhomogeneous distributions of carrier densities when this density at the metal/semiconductor boundary deviates from the density in the semiconductor bulk. As a consequence, a space charge layer is formed which creates inhomogeneous field and potential distributions near the electrode. When the deviation of the boundary density from the bulk density is significant, it influences the current through the semiconductor, reducing it by creating a Schottky barrier when $n_c \ll n_{10}$, and increasing it via carrier injection when $n_c \gg n_{10}$.

A different kind of inhomogeneous distribution of the electric field may occur in the regime of *nonlinear transport far from thermodynamic equilibrium*. In that case, the spatial patterns arise in a self-organized way, even in homogeneous semiconductors. For more on such self-organization, see Chap. 5. In Sect. 2.1, we will analyze stationary high-field domains that must occur in homogeneous semiconductors when the conductivity decreases stronger than linearly with the electric field. For this, we will discuss the system of transport and Poisson equations in a one-dimensional n-type semiconductor with the help of a field-of-direction analysis also known as a phase portrait analysis. This will give conclusive information about the type of possible solution curves of these equations, hence will assist in defining under what conditions high-field domains must occur.

2.1 Field-of-Direction Analysis

This field-of-direction analysis is suited to obtain general information on the type of solution curves of the set of transport and Poisson equations easily and can be used to obtain quantitative information in a simple but satisfactory approximation. This analysis can be used easily in a *one-carrier model* of a one-dimensional semiconductor in which the set of governing equations (transport, Poisson, and continuity equations)

$$j = e\mu_n nF + \frac{kT}{e}\frac{\mathrm{d}n}{\mathrm{d}x} \tag{2.1}$$

and

$$\frac{\mathrm{d}F}{\mathrm{d}x} = \frac{e}{\varepsilon\varepsilon_0}(p_\mathrm{d} - n - n_\mathrm{a}) \tag{2.2}$$

may be re-written as a set of two differential equations[1]

$$\frac{\mathrm{d}n}{\mathrm{d}x} = \frac{e}{kT}\left(\frac{j}{e\mu_n} - nF\right) \tag{2.3}$$

and

$$\frac{\mathrm{d}F}{\mathrm{d}x} = \frac{e}{\varepsilon\varepsilon_0}\left(p_\mathrm{d} - n - n_\mathrm{a}\right). \tag{2.4}$$

For definiteness we will assume first that the field dependence enters via reaction kinetics and makes the carrier density n, the density of electrons in acceptors n_a, and of holes in donors p_d implicit functions of F.

The solution curves of interest are $n(x)$ and $F(x)$, and, derived from the latter, the electron potential distribution $\psi_n(x)$. Since (2.3) and (2.4) are an autonomous system of differential equations that does not contain x explicitly, and fulfills the Lipschitz conditions of having continuous partial derivatives with respect to F and n, the solution curves can be projected into any nF plane, yielding a single-valued curve. That is, through every point of the nF plane traverses one and only one solution curve; or, every point in this plane uniquely defines *one* slope, except for *singular points* or *fixed points* which are defined by

$$\frac{\mathrm{d}F}{\mathrm{d}x} = \frac{\mathrm{d}n}{\mathrm{d}x} \equiv 0. \tag{2.5}$$

Solutions cannot leave or enter singular points which represent solutions in which $n(x) = \text{const}$ and $F(x) = \text{const}$. One also concludes that no solutions can cross each other in the entire nF plane since each of its points presents one direction only.

When identifying the direction of increasing x from the cathode toward the anode, one can attach to each point in the nF plane an arrow indicating this direction (Fig. 2.1). This field of arrows is known as a *field of directions.* When the boundary values are given ($n(x = 0)$ and $F(x = 0)$), one can start

[1] For an n-type semiconductor, and assuming $p \ll n$ and $j_p \ll j_n \simeq j$.

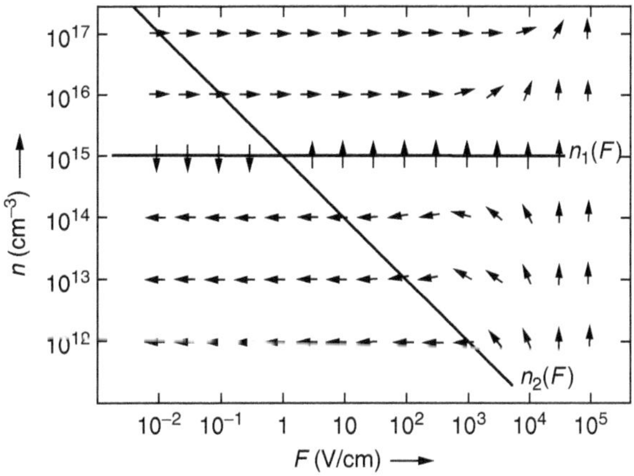

Fig. 2.1. Field of directions computed for a typical homogeneous semiconductor

from this point and follow along the directions from point to point to obtain the solution curve $n(F)$. Since an x-value is attached to every point in the nF plane, one finally obtains the desired $n(x)$ and $F(x)$. Such a procedure seems to be tedious.

However, with the help of two auxiliary curves which can easily be identified, a simple method can be derived, which permits immediate classification of the possible set of solution curves. These auxiliary curves (*null-isoclines*) are those for which Böer and Wilhelm (1963)

$$\frac{\mathrm{d}F}{\mathrm{d}x} \equiv 0, \qquad \text{identified as} \qquad n_1(F) \tag{2.6}$$

and

$$\frac{\mathrm{d}n}{\mathrm{d}x} \equiv 0, \qquad \text{identified as} \qquad n_2(F), \tag{2.7}$$

and are shown in Fig. 2.2.

The first curve, $n_1(F)$, is known as *quasi-neutrality curve*, since, following the Poisson equation, the space charge vanishes for any point on this curve. It can be crossed only vertically in the nF plane since, according to (2.6) F remains constant with changing x. If F changes in x, the $n_1(F)$ curve is the locus of field extrema.

The second curve, $n_2(F)$, is known as the *drift current curve* (see (2.3)), since for any point on this curve the diffusion current vanishes. It is therefore also referred to as a *diffusion–neutrality curve*. This curve can only be crossed horizontally in the nF plane since n remains constant with changing x according to (2.7).

Below $n_1(F)$ one obtains $\mathrm{d}F/\mathrm{d}x < 0$;[2] above $n_1(F)$, one has $\mathrm{d}F/\mathrm{d}x > 0$, as seen from (2.4). In a similar fashion one sees that $\mathrm{d}n/\mathrm{d}x < 0$ below and >0

[2] Observe that the field F is negative here.

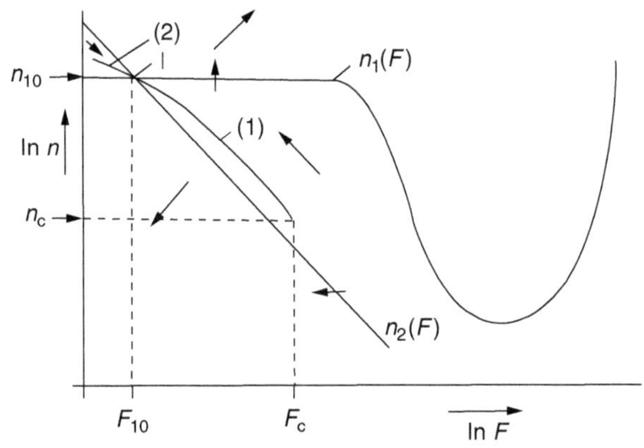

Fig. 2.2. Symbolized field of directions with quasi-neutrality curve ($n_1(F)$) and drift current curve ($n_2(F)$); *solid arrows* indicate the four quadrants of possible direction of the solution curve. Such solution curves are shown as curves (1) and (2)

above $n_2(F)$. Therefore, the direction of any solution curve $n(F)$ points into different quadrants below or above the quasi-neutrality and the drift current curves, as symbolized by the four heavy arrows[3] in Fig. 2.2. This figure also shows typical shapes for the $n_1(F)$ and $n_2(F)$ curves in semiconductors that have a regime of negative differential conductivity followed by a regime in which the conductivity increases again with the field.[4]

The $n_1(F)$ curve shows the field dependence of the electron density under the neutrality condition; i.e., in the given one-carrier model, $n_1(F)$ is the same as $n(F)$ in a homogeneous, space-charge-free semiconductor. It is constant for low fields and usually increases at sufficiently high fields because of field excitation.

To show a more complete example, we have chosen a photoconducting CdS that with optical carrier generation may have an intermediate field range[5] in which n decreases with increasing F because of field quenching (Dussel and Böer 1970). This decreasing branch identifies the range of negative differential conductivity and is included in the schematically drawn example given in Fig. 2.2. For quantitative descriptions, we refer to further sections in this book.

---

[3] The arrow in the upper right quadrant, e.g., indicates that through any point in this quadrant the solution curve must have a direction between 0° and 90°; this holds correspondingly for the other quadrants.

[4] This is typical for some copper-doped CdS crystals in which the electron density with increasing fields first decreases (field quenching – see later in the text) and then increases because of field excitation.

[5] This intermediate field range is characterized by Frenkel–Poole excitation of carriers from Coulomb-attractive centers that can start at fields as low as a few kilovolts per centimeter, as opposed to the much higher fields necessary for ordinary field excitation from neutral centers that typically need fields in the 100 kV/cm range.

The $n_2(F)$ curve is a hyperbola for a given j, and as long as the mobility is field-independent. In the double logarithmic representation of Fig. 2.2, it is given as a straight line with a slope of -1. The crossing point of curves $n_1(F)$ and $n_2(F)$ is the singular point I. In the bulk of a semiconductor, which is usually several orders of magnitude thicker than the Debye length, $n(x)$ and $F(x)$ change very little with x; hence the solution curve must "approach" the singular point. It should, however, be remembered that no solution can leave or cross a singular point. Thus, a physically meaningful solution will always start or end in one of the quadrants, but slightly removed from the singular point.

A typical solution curve for a Schottky barrier is shown as curve 1 in Fig. 2.2. It starts at the metal/semiconductor interface ($x = 0$) at a lower density n_c and a higher field F_c than in the bulk (given by the singular point) and follows the field of directions (see Fig. 2.1) toward the singular point I.

Another solution is shown for an injecting contact as curve 2 in Fig. 2.2. It starts at a higher electron density and consequently at a lower field at $x = 0$, and also moves toward the singular point I.

One value of the field-of-direction analysis lies in the fact that one can immediately identify what types of solutions are possible. These are given as curves 1 and 2 and are the only types possible as stationary solutions for a thick (compared to the Debye length) device.[6]

From Fig. 2.2, one sees that it is possible to approach the singular point I from the second or fourth quadrants only; it is impossible to enter any other quadrants from the first or third quadrant. Namely, crossing the $n_1(F)$ or $n_2(F)$ curve from the second or fourth quadrant means irrevocably turning away from the singular point, yielding physically meaningless solutions. In mathematical terms, the singular point I is a saddle point (see Sect. 3.2), i.e., it is impossible to circle it.

In Fig. 2.3, a composite display shows the relation of $n(x)$, $F(x)$, and $n(F)$ in the field of directions representation for the Schottky barrier. It also gives the corresponding band diagram in Panel 2.3d.

Cathode-Adjacent and Anode-Adjacent Domains

When increasing the bias to the device, the current increases, thereby causing a parallel shift of the drift current curve toward higher values, e.g., from curve (1) to curve (2) in Fig. 2.4. If the field-quenching branch of the quasi-neutrality curve decreases more than linearly with the field, then $n_1(F)$ and $n_2(F)$ will cross again, thereby creating two more singular points (II and III in Fig. 2.4).

[6] These solutions represent a device with one electrode. When including the other electrode one must be careful with the signs in the transport equations which result in a reversal of the arrows in the field of directions when approaching the second electrode; this reversal permits the traverse of the projection of the solution curve in the opposite direction in part of the solution.

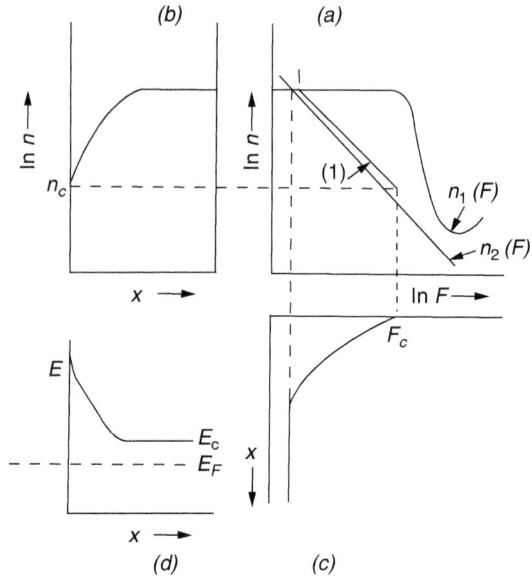

Fig. 2.3. Composite schematic representation of a Schottky-barrier device. (**a**) Field of directions with quasi-neutrality curve (n_1) and drift current curve (n_2); solution curve 1 starting at (n_c, F_c) and approaching the singular point I; (**b**) the corresponding $n(x)$; (**c**) the corresponding $F(x)$; and (**d**) the corresponding band diagram with conduction band E_c and Fermi level E_F

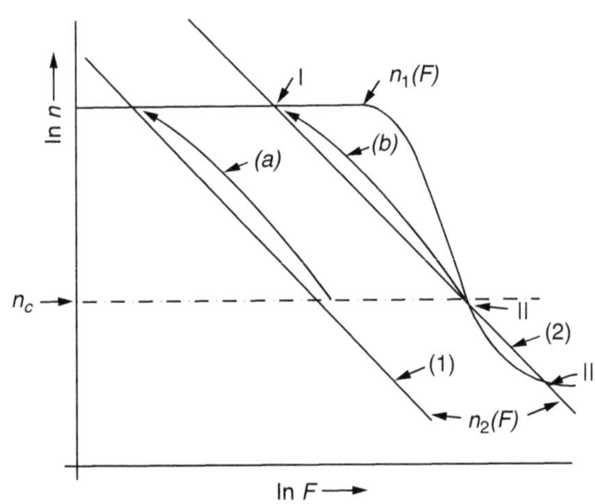

Fig. 2.4. Field of directions representation with quasi-neutrality curve (n_1) and two drift current curves (n_2) for lower (1) and higher (2) current with corresponding solution curves (**a**) and (**b**)

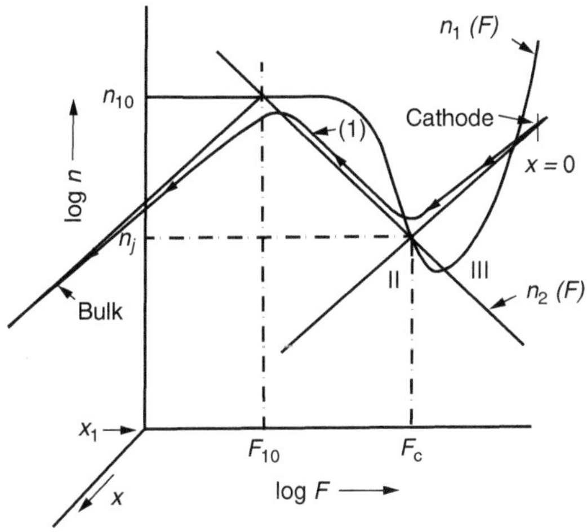

Fig. 2.5. Three-dimensional representation of a solution curve (1) extending between two singular points (II and I) and showing a step like character in n and F; $x = 0$ indicates the position of the cathode, $x = x_1$ the position of the step edge

Assuming for simplicity that the boundary density n_j does not change with increasing bias and current ($n_j = n_c$), the field at the boundary increases until the starting point of the solution has moved close to the second singular point. Then, the drift current curve cannot rise further because it would force the solution curve to cross $n_1(F)$ while maintaining the boundary condition at n_c. Such a crossing of $n_1(F)$ from the fourth quadrant is impossible. This means that with further increasing bias $n_1(F)$ can no longer rise, the *current must saturate* and the solution expands in x near the second singular point. This solution (2) becomes *step like*, as shown in Fig. 2.5 in a three-dimensional representation; i.e., $n(x)$ and $F(x)$ are almost constant near $x = 0$, rather than $n(x)$ steeply increasing and $F(x)$ linearly decreasing near $x = 0$ in a Schottky barrier. This distinctly different, step-like behavior of $n(x)$ and $F(x)$ is referred to as the *high-field domain* (Böer et al. 1959, 1963).

With increased bias, the high-field domain widens without increasing the field at the cathode (in n-type materials), while the current remains constant. In the field of directions representation, no changes are visible, except for a minute extension of the solution closer to the singular point II.

As long as the field remains constant within the domain, here also the space charge vanishes. When the field is increased to initiate field quenching, localized holes in deep hole traps, that produce this space charge in compensated photoconductors, are freed and can either move into the adjacent electrode or recombine, thus reducing the positive space charge. With sufficient depletion of these hole traps, the space charge is reduced to zero.[7]

[7] Overcompensation is avoided since a reversal of space charge would reduce the field that was the cause for field quenching, and automatically cause an adjustment to yield $\varrho \equiv 0$.

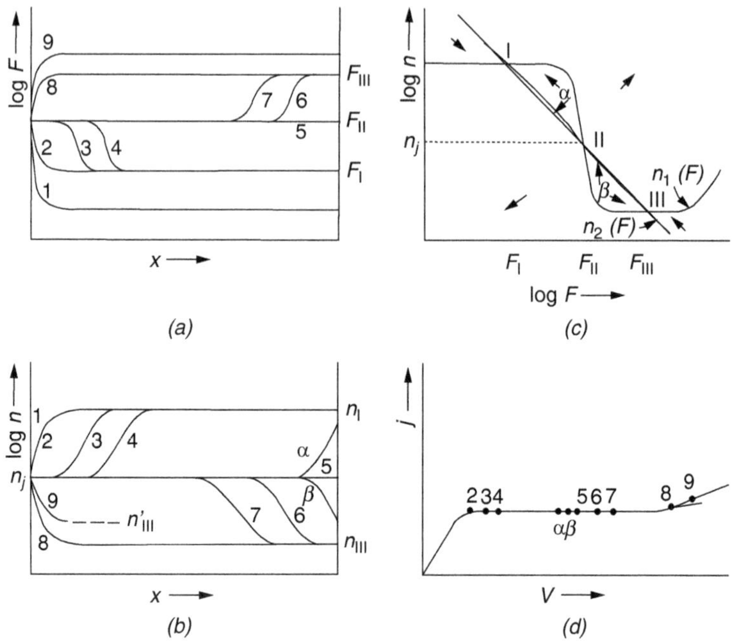

Fig. 2.6. (**a**) Field distribution in high-field domains with bias as family parameter; (**b**) corresponding electron density distribution; (**c**) field of directions with solution curves (α) for cathode-adjacent and (β) for anode-adjacent high-field domains; (**d**) corresponding current–voltage characteristic with the bias identified by numbers that are used as family parameter for sub figures (**a**) and (**b**)

When the bias is increased sufficiently so that the high-field domain has expanded to fill the entire device, the solution remains at the singular point II. A further increase in bias causes a minute increase of the current (a slight shift upward of $n_2(F)$) that brings the solution into the quadrant between the singular points II and III, and between $n_1(F)$ and $n_2(F)$; it extends the solution from II toward the singular point III. A still further increased bias causes part of the solution (toward the anode) to approach III, now forming an anode-adjacent domain of even higher fields ($F_{\rm III}$), which then expands from the anode toward the cathode until it fills the entire device (see Fig. 2.6c).

A family of these field distributions is shown (schematically) in Fig. 2.6a and with electron density distribution above and below the singular point shown in Panel b for cathode- and anode-adjacent high-field domains, and for intermediate and high bias, respectively. The corresponding points on the current–voltage characteristics are shown and numbered accordingly in Panel d.

At a still higher bias, when most of the solution lies near the singular point III (curve 8), the current must again increase to maintain the basic boundary condition $n(x = 0) = n_c$. This causes an increase of the field in the anode-adjacent domain to absorb the additional bias. The field distribution has turned from a Schottky barrier type (curve 1) to an injecting contact type of solution (curve 9) by a swing of the electron density with increasing distance from the cathode from above to below n_c. A further increased bias causes the current and thus the singular point to shift upward, making the contact neutral when $n_{III} = n_c$ and finally, for $n_{III} > n_c$, again producing a Schottky contact.[8]

In Sect. 2.1, we will show that in CdS a major part of this domain behavior can be experimentally confirmed before electric breakdown phenomena set in. Without a field-of-direction analysis this domain behavior would be rather difficult to explain. The given discussion, however, even though qualitative, makes such analysis transparent. It is for this reason that we have included the more qualitative field of directions method with the otherwise preferred numerical or analytical discussion.

We will also show in further sections that in a number of realistic cases this method provides enough information to yield unambiguously the carrier density and mobility as a function of the electric field from rather simple experiments. The key to the success of this simple analysis is the fact that the field is essentially constant within a domain, and the transition region between bulk and domain has a thickness of only a few Debye lengths, i.e., its width can usually be neglected compared to the thickness of the device and the domain. Then, a simple step approximation is sufficient to describe the voltage drop across the device:

$$V = F_{II}x_1 + F_I(d_1 - x_1). \tag{2.8}$$

That is, with changing bias the width x_1 of the high-field domain changes linearly. This permits the determination of F_I and F_{II} when the domain width is measured as function of the bias.

Franz–Keldysh Effect to Directly Observe High-Field Domains

The domain width can be measured directly with probing light, using the Franz–Keldysh effect (Franz 1958; Keldysh 1958 – see Böer 1959), which causes the absorption edge to shift toward longer wavelengths with increasing field. When observing the light transmitted through the semiconductor platelet,[9] one detects the domain as region of increased absorption when using light at the absorption edge (Böer et al. 1958, 1959; Böer 1961), as shown in Fig. 2.7.

---

8 In this discussion, we have neglected drift velocity saturation (see Böer 2009). When such saturation is included, the injection range vanishes. We have also neglected a field-dependence of n_c.

9 For reasons of optical observation, the geometry of the device was chosen so that the thin platelet dimension is normal to the vector of the probing light and the field direction lies within it and is perpendicular to it.

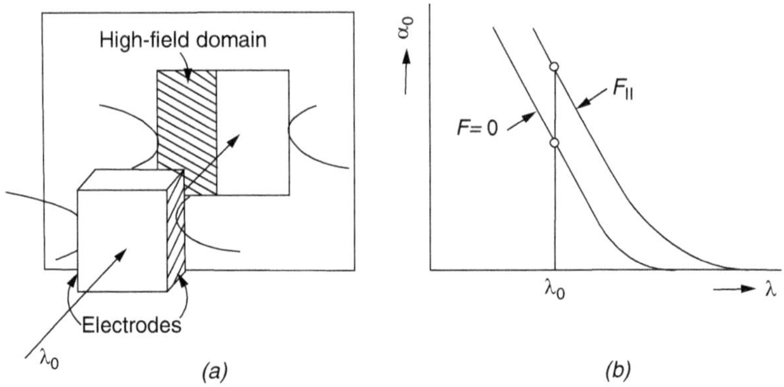

Fig. 2.7. (**a**) Experimental setup to use the Franz–Keldysh effect for observing high-field domains. (**b**) Shift of the absorption edge due to the electric field. The crystal platelet is viewed in transmitted monochromatic light (λ_0) that shows the high-field domain as a region of increased absorption (*shadow*)

Such high-field domains were first seen in CdS (Böer et al. 1958) and have since been extensively investigated with this method.

The high-field domains are directly visible in the shadow projection given in Fig. 2.7 and show[10] in copper-doped CdS[11] the widening of the cathode-adjacent domain with increasing bias until the domain fills the entire device at 1,750 V (upper row of Fig. 2.8a). With further increased bias, an anode-adjacent domain of higher field strength (with an increased band-edge shift, hence darker shadow) emerges from the anode and broadens with further increased bias (lower row of Fig. 2.8a). The current–voltage characteristic shows no break in the saturation behavior during the transition from cathode- to anode-adjacent domain (indicated by the arrow in Fig. 2.8b), in agreement with the prediction from the field-of-direction analysis (Böer and Voss 1968).

When plotting the domain width as a function of the bias for the cathode-adjacent (a) and anode-adjacent (b) high-field domains, as shown in Fig. 2.9a, one obtains from the slopes and intersections with the abscissa according to (2.8) the fields of the three singular points. These are shown in Fig. 2.9b. The electron densities then are directly obtained from the drift current curve.

2.1.1 Electron Mobility as Function of the Field

The stationary high-field domain gives a unique opportunity to determine the electron mobility unambiguously as a function of the electric field. Since the electric field throughout an entire high-field domain is constant, this is the only

[10] Observe the broadening gray band adjacent to the left electrode and at a bias >2,000 V an even darker gray band adjacent to the right electrode. Disregard the fine wavelike pattern caused by crystal defects ("striations").

[11] A similar behavior was also obtained in some Si samples (Böer and Williges 1961).

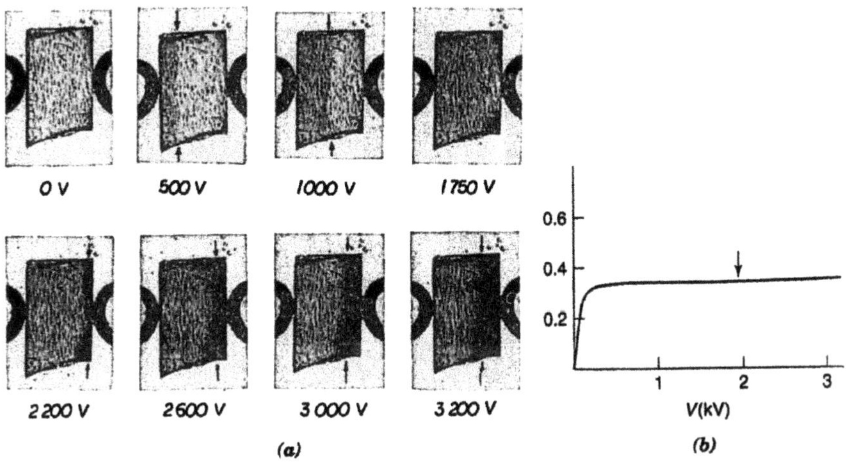

Fig. 2.8. (**a**) Photographs of the CdS crystal shown between the two electrodes. The "waves" within the crystal are caused by crystallographic striations and their appearance should be neglected. In the upper row a slight darkening is observed extending from the left electrode (cathode) and increasing in width with increasing applied voltage as indicated below each photo the increasing width is indicated by the two appears above and below each photograph. In the lower row of photos the darkening shown in the upper row has extended to the anode, and a still stronger darkening is now extending from the anode that increases in width with further increasing applied voltage. (**b**) Current–voltage characteristic corresponding to the results given in (**a**) with the *arrow* pointing to the transition between cathode and anode adjacent domain (After Böer and Voss 1968)

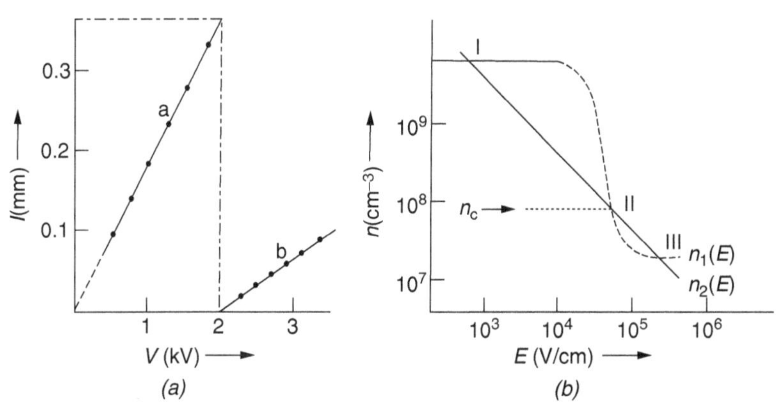

Fig. 2.9. (**a**) Domain width as function of bias in CdS at 230 K, obtained from the experiment described in Fig. 2.8a. (**b**) Singular points obtained from the slopes of the curves a and b in Panel (**a**) (After Böer and Voss 1968)

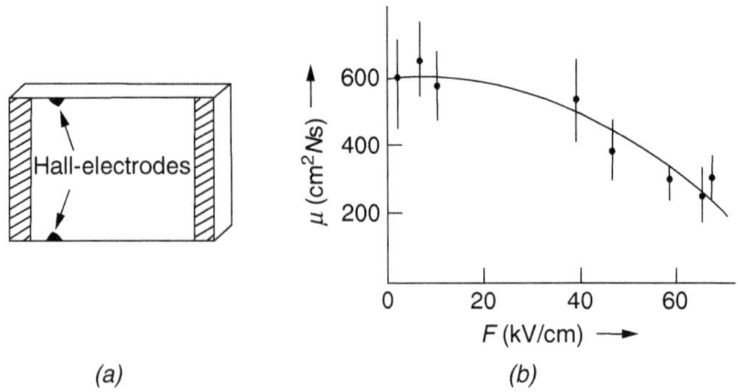

Fig. 2.10. (**a**) Electrode arrangement for measuring the Hall effect within a high-field domain. (**b**) Hall mobility in CdS as function of the electric field measured in a high-field domain at 230 K (After Böer and Bogus 1968)

opportunity to confirm the value of the actual field.[12] The field-dependent Hall mobility was measured using high-field domains (Böer and Bogus 1968) with Hall electrodes close to the cathode and applying a sufficient bias so that the cathode-adjacent domain extends well beyond the Hall electrodes. The low-field mobility can be checked by reversing the polarity, thereby placing the Hall probes in the low-field bulk. Different pairs of experimental points were obtained by using different metal electrodes yielding different values of the boundary electron density n_c, hence shifting the neutrality curve and thereby changing the position of the singular points and thereby the field within the high-field domain.

The results of these experiments are shown in Fig. 2.10. Starting near 20 kV/cm, the mobility decreases with increasing field. Such a decrease follows a $1/F$ behavior at higher fields (see Böer 2009) and causes a saturation of the drift velocity ($\mu F = \text{const}$) as it approaches the rms-velocity of the electrons.

2.1.2 Drift Velocity Saturation at Higher Fields

The drift velocity of electrons increases linearly with the acting electric field according to $v_d = \mu F$. However, when, at sufficiently high electric fields, the mobility decreases with field linearly, the drift velocity no longer increases but saturates. This has significant influence on the field of direction in a bending of the drift current curve $[n_2(F)]$ and results in corrected values for the first and third singular point as shown in (Fig. 2.11). The figure also reproduces as dash-dotted curves the results given in Fig. 2.9b for comparison.[13]

[12] Such electric field is otherwise only estimated from simple geometric assumptions, using applied voltage and electrode distance and neglecting any possible field inhomogeneities, caused by electrode or crystal defects.

[13] The shift of the first singular point is not resolved experimentally because of the limited accuracy of the method to determine this point.

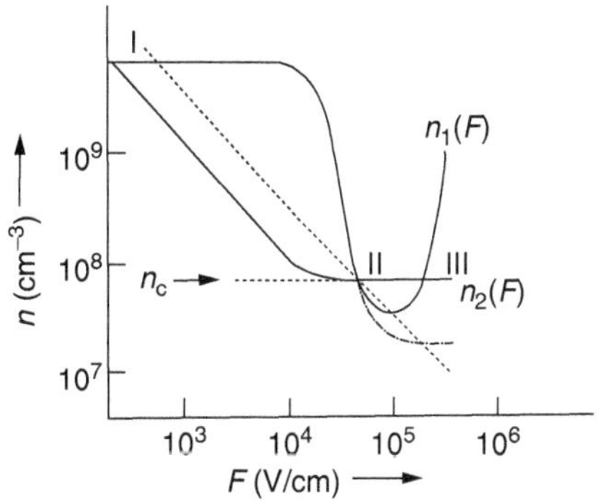

Fig. 2.11. Field of directions with corrected $n_2(F)$ and values for the three singular points obtained by the corrected experimental values. Uncorrected $n_1(F)$ and $n_2(F)$ shown as *dashed curve* segments

Incomplete Drift Saturation

The field of directions requires that for complete drift velocity saturation, the third singular point can only be reached when the solution curve proceeds directly on the drift current curve. Even a slight deviation below this curve would force the solution away from the third singular point (see the arrows in Fig. 2.6c). In actuality, however, the drift velocity saturation is incomplete when an external field is acting; in this range, the drift velocity is not equal to, but approaches with increasing field the thermal rms velocity of carriers. The $n_2(F)$ curve remains, therefore, slightly slanted, permitting a deviation of the solution curve from the vicinity of II below this drift current curve (i.e., proceeding between $n_1(F)$ and $n_2(F)$) and still reaching the third singular point.

The described method has the advantage of providing an unambiguous means to determine the field and consequently the electron densities at these fields. It has the disadvantage of identifying only three fixed values of the field and electron density for a given sample: $F_{\mathrm{I}}, n_{\mathrm{I}}$; $F_{\mathrm{II}}, n_{\mathrm{II}}$; and $F_{\mathrm{III}}, n_{\mathrm{III}}$. To vary the field of a high-field domain, we will now describe an example of a more advanced method, using the field-of-direction analysis.

2.2 Pseudo-Cathodes and the Field of Directions

The quasi-neutrality curve shown in the previous sections was drawn by interpolation between only three measured points, the singular points at F_{I}, F_{II}, and F_{III}. The important second singular point for this interpolation is determined by the intersection of the quasi-neutrality curve $n_1(F)$ with the electron density n_c at the metal/semiconductor boundary. When this density is changed over a large range, almost the entire $n_1(F)$ curve in the important range of negative differential conductivity can be obtained. However, rather than using different metal contacts with various work functions,[14] one can employ a pseudo-cathode in photoconductors, which serves a similar purpose with respect to a change of the boundary density.

Such a pseudo-cathode can be created by a region of lower illumination close to the cathode, which is wide enough[15] to screen the influence of the actual cathode. The electron density is reduced in this region and its edge acts as a quasi-cathode with an easily variable boundary density n_j, using various intensities of illumination in this region. Experimentally, this is accomplished by placing a thin shadow band close to the metal cathode and providing a variable, low-intensity light within this band (see Fig. 2.12a).

The solution curves with such pseudo-cathodes can be discussed in a more complex composite field of directions diagram that is shown in Fig. 2.13, however follows the same roles for the solution curve analysis in the simple diagram shown before. This diagram contains two $n_1(F)$ curves; the original curve and the one in the shadow band that is identified by $n_s(F)$. Assuming that the cathode is slightly blocking, the solution curve $n(F)$ starts at the cathode density that is smaller than in the adjacent part of the bulk: $n_c < n_1(F = 0)$, hence produces a Schottky barrier that moves along the branch P until it approaches the singular point I (this part was not shown in the previous discussion and was neglected there since it was small compared to the domain). Then, the solution curve proceeds after reversing its direction before it enters the region of the shadow band. When the solution then enters the shadow band, it is now above $n_s(F)$ where the field of directions reverses its sign and points in the fourth quadrant (as shown by the dashed arrow). The solution therefore can move toward lower n and higher F values (as given by the dashed branch Q), and with low-applied voltages approach the singular point I', assuming that the shadow region is much wider than a few Debye lengths (Fig. 2.13a). Then, it crosses $n_2(F)$ and moves again up to the first singular point (branch R) much like from another Schottky barrier. All of this is not visible since it occurs within the shadow region. After passing through the main bulk (now above n_2), it then reaches the second electrode (at the arrow head in Fig. 2.13a) that is assumed here to be of the same metal (hence has the same n_c).

When the applied voltage is increased, the drift current curve $n_2(F)$ is shifted up until it loses the intersection with $n_s(F)$ in the shadow (Fig. 2.13b). With still close proximity between $n_2(F)$ and $n_s(F)$, the dashed part of the solution (branch Q') can squeeze through a the high part of this saddle point ($\mathrm{I}^{\times}$) providing a long enough path within the shadow, then, near the edge of the shadow, goes on to approach the domain-singular point II. Near the edge of the shadow, the electron density drops and the field rises until n_{II}, F_{II} is

---

[14] This means that one has to use different individual crystals that not necessarily have the same doping and defect structure and thereby introduce some ambiguity in the interpretation of the results.

[15] That is several Debye length wide.

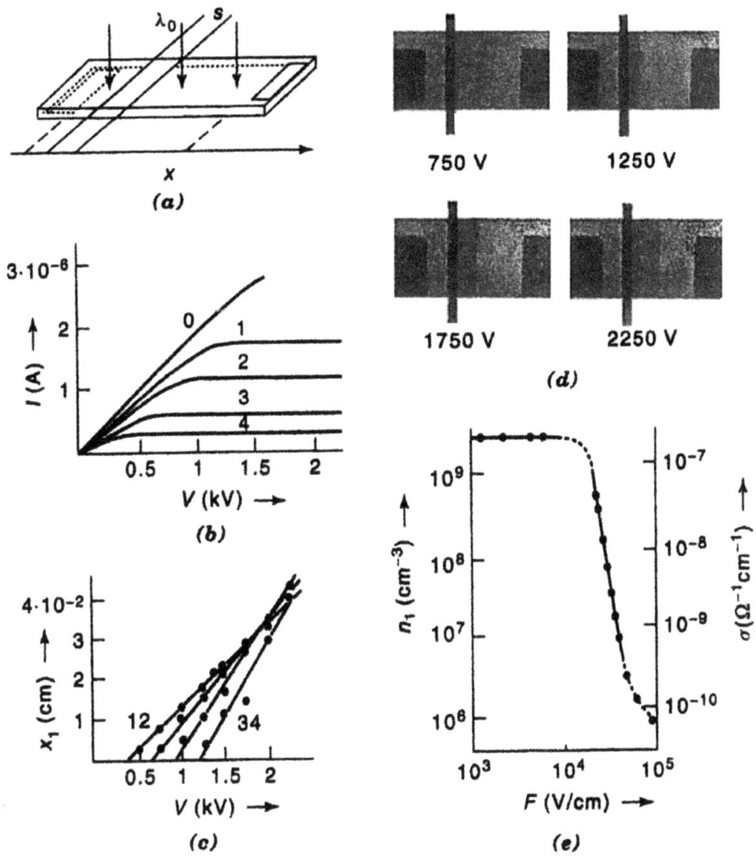

Fig. 2.12. (**a**) Pseudo-cathode experimental setup with shadow band close to cathode. (**b**) Current–voltage characteristics without (0) and with shadow and diminishing light in the shadow band causing a decrease of the saturation current (with increased shadow density from curves 1 to 4). (**c**) Domain width as function of bias (same family as in **b**). (**d**) Photographs of the CdS crystal between the two electrodes with a shadow band extending beyond the crystal, as shown in (a). The slight shadow extending to the right from the shadow band shows the high-field domain that increases in width with increasing applied voltage. (**e**) Measured quasi-neutrality curve using this pseudo-cathode experiment (After Böer et al. 1968)

almost reached. The solution then crosses $n_2(F)$ and continues in the fashion described in Sect. 2.1.2 for a typical high-field domain that now appears adjacent to the pseudo-cathode. Even though complex to follow in detail, the effect of the shadow band is that it indeed acts as a pseudo-cathode from which at sufficiently high-applied voltage a high-field domain can be seen to extend beyond the shadow by its Franz–Keldysh shift of the absorption edge.

Such high-field domains adjacent to the pseudo-cathode are shown in Fig. 2.12d and widen with increased bias (Panel c), permitting the determination of $(n_{\rm II},\ F_{\rm II})$. After changing the light intensity in the shadow band,

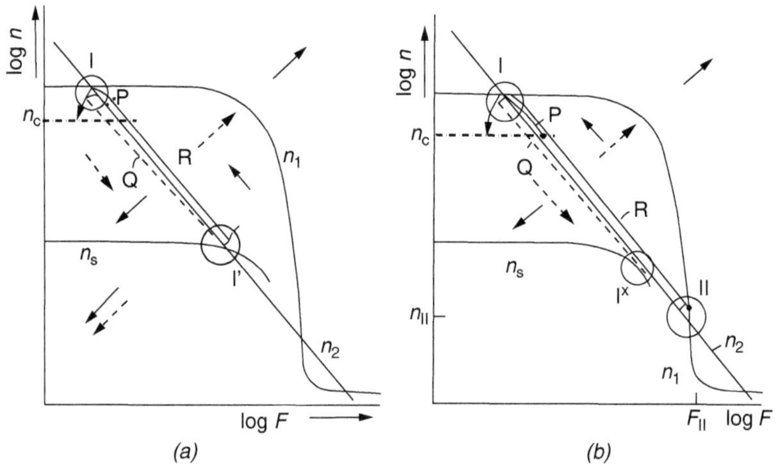

Fig. 2.13. Field of directions for a long photoconductor sample with pseudo-cathode. Solution curves shown in cathode-adjacent bulk (branch P), shadow band of pseudo-cathode (*dashed* branch Q with corresponding arrow symbols to indicate direction quadrant), and main device bulk (branch R). (**a**) for low voltage applied; (**b**) for higher voltage, resulting in a high-field domain near the singular point II

another set of (n_{II}, F_{II}) is obtained, etc., until the declining branch of the $n_1(F)$ curve is obtained (Panel e).

When the pseudo-cathode shadow band is positioned close enough to the anode, the cathode-adjacent domain can be extended up to the anode without requiring excessive applied voltages. With further increased bias, anode-adjacent domains can then be obtained, thus also probing the increasing branch of the $n_1(F)$ curve (Fig. 2.14).

This method has been used to more closely study the mechanism that causes the decreasing density of photo-generated electrons (field quenching (Dussel and Böer 1970)), and the transition between stationary and moving high-field domains, that will be discussed in Sect. 5.3.

In contrast, the application of metals with different work functions for analyzing the $n_1(F)$ behavior is also hampered by the bias dependence of the electron density at the metal/semiconductor contact (Böer et al. 1969; Stirn et al. 1971; Dussel et al. 1973; Stirn et al. 1973). This is discussed in Sect. 2.2.1.

2.2.1 Doping-dependence of the Field Quenching

We first, however, will discuss another information that we can obtain from high-field domains with the method just described. The negative differential conductivity in copper-doped CdS is caused by field quenching due to Frenkel–Poole excitation of Coulomb-attractive hole traps that are the result of copper doping. The field dependence of the carrier density can be estimated from a reaction kinetic model that shows the relevant transitions between the bands

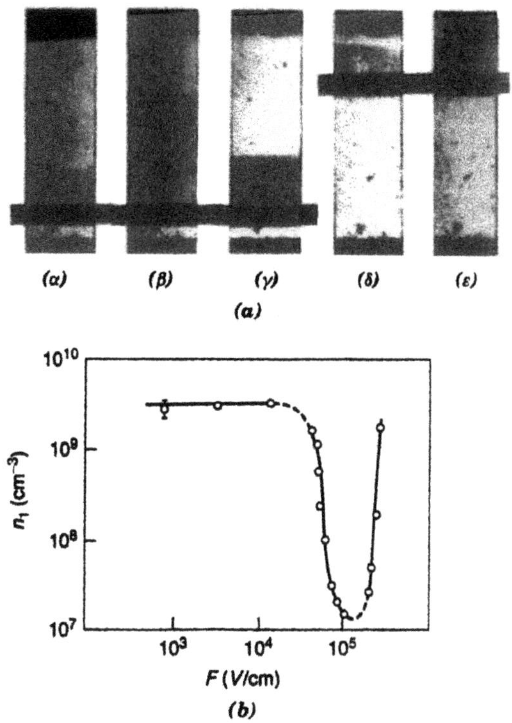

Fig. 2.14. (**a**) CdS crystal shown between the cathode (below) and the anode (above) the crystal. A shadow band extending beyond the crystal is shown in the lower half of the crystal in the first three photos and in the upper part in the last two photos. (α, β) with increasing bias causing a widening of the high-field domain; (γ) with reduced light intensity in shadow causing a higher domain field seen by increased darkening; (δ) with shifted pseudo-cathode; and (ϵ) with increased bias to produce an anode-adjacent domain. (**b**) Measured quasi-neutrality curve $n_1(F)$ using this pseudo-cathode method (After Böer 1969)

and the trap and recombination levels shown in Fig. 2.15. The corresponding carrier densities can be determined from the kinetic equations considering all transitions:

$$\frac{\mathrm{d}n}{\mathrm{d}t} = g_o + c_{\mathrm{tc}} n_{\mathrm{t}} N_{\mathrm{c}} - n c_{\mathrm{ct}} (N_{\mathrm{t}} - n_{\mathrm{t}}) - n c_{\mathrm{rs}} p_{\mathrm{s}} - n c_{\mathrm{rf}} p_{\mathrm{f}}$$

$$\frac{\mathrm{d}n_t}{\mathrm{d}t} = n c_{\mathrm{ct}} (N_{\mathrm{t}} - n_{\mathrm{t}}) - N_{\mathrm{c}} c_{\mathrm{tc}} n_{\mathrm{t}}$$

$$\frac{\mathrm{d}p_{\mathrm{s}}}{\mathrm{d}t} = p c_{\mathrm{vs}} (N_{\mathrm{s}} - p_{\mathrm{s}}) - n c_{\mathrm{cs}} (n - p_{\mathrm{s}}) - p_{\mathrm{s}} c_{\mathrm{sv}} (F)(N_{\mathrm{v}} - p_{\mathrm{s}})$$

$$\frac{\mathrm{d}p_{\mathrm{f}}}{\mathrm{d}t} = p c_{\mathrm{vf}} (N_{\mathrm{f}} - p_{\mathrm{f}}) - n c_{\mathrm{cf}} (n - p_{\mathrm{f}})$$

$$\frac{\mathrm{d}p}{\mathrm{d}t} = d_{\mathrm{s}} c_{\mathrm{sv}} (N_{\mathrm{v}} - p_{\mathrm{s}}) - p_{\mathrm{f}} c_{\mathrm{fv}} (N_{\mathrm{v}} - p_{\mathrm{f}}) + p_{\mathrm{c}} v_{\mathrm{s}} (F)(N_{\mathrm{v}} - p_{\mathrm{s}})$$

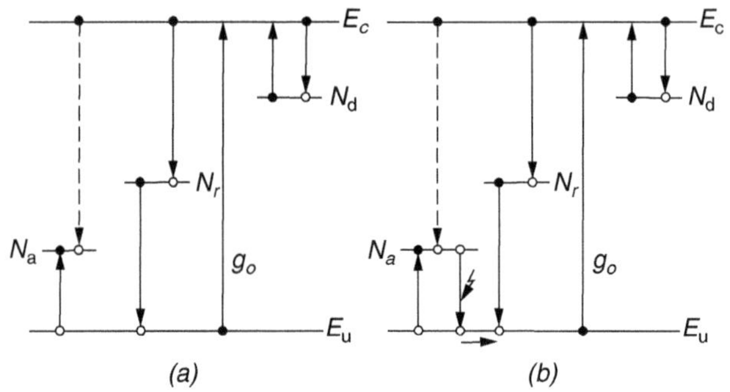

Fig. 2.15. Reaction Kinetic model of CdS with three types of defect centers identified by indices a, d, and r for acceptor, donor and recombination center and electron transitions identified by arrows (**a**). A redistribution of carriers away from acceptors is indicated by a *field-arrow* in (**b**)

with c_{ik} identifying transition coefficients from level i to k and small letters identifying carrier densities and capital letters level densities. These kinetic equations turn into polynomial equations when in steady state all time derivatives vanish. Field quenching is described by the field excitation $c_{\mathrm{vs}}(F)$ of trapped holes from slow copper recombination centers (acceptors) that can then be recaptured by fast recombination centers and force the optically generated electron density to decrease by enhanced recombination through these fast recombination centers. The field excitation of trapped holes from these slow Coulomb attractive acceptor-like (to holes, hence repulsive to electrons) recombination centers into the valence band occurs because of Frenkel Poole excitation.

Such Frenkel Poole excitation of Coulomb attractive shallow defect centers occurs at relatively low electric fields, typically in the range of 20–50 kV/cm that is far below other field excitation mechanisms such as tunneling or impact ionization start.

A simple potential diagram shown for a Coulomb attractive electron trap is shown in Fig. 2.16 that indicates a substantial lowering of the funnel that extends typically to 100 Å by more than kT.

A similar lowering of the excitation energy takes place for Coulomb attractive hole traps. If this assumption is correct, however, one expects that the degree of field quenching should decrease with decreasing distance between these Coulomb attractive traps when their Coulomb funnels start to overlap significantly. This decrease of the Frenkel–Poole excitation is expected when the Coulomb attractive traps approach each other to within less than 100 Å. Since these centers are generated by copper doping, it is expected that with increasing copper doping the effectiveness of the lowering of the Coulomb barrier would be reduced (see Fig. 2.17).

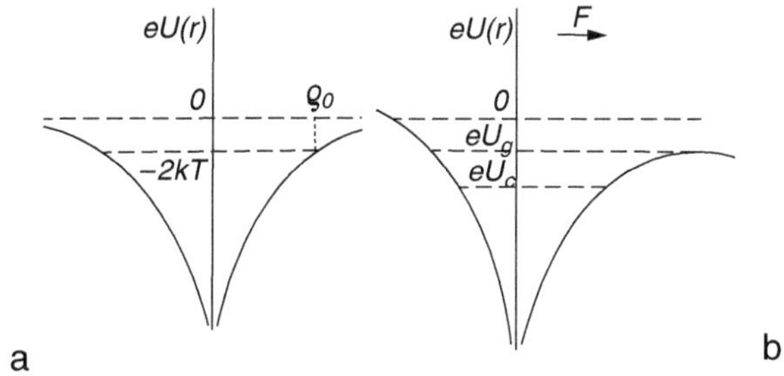

Fig. 2.16. Potential funnel of a Coulomb attractive electron trap (**a**) without and (**b**) with an acting field, indicating the lowering of the excitation energy from the trap

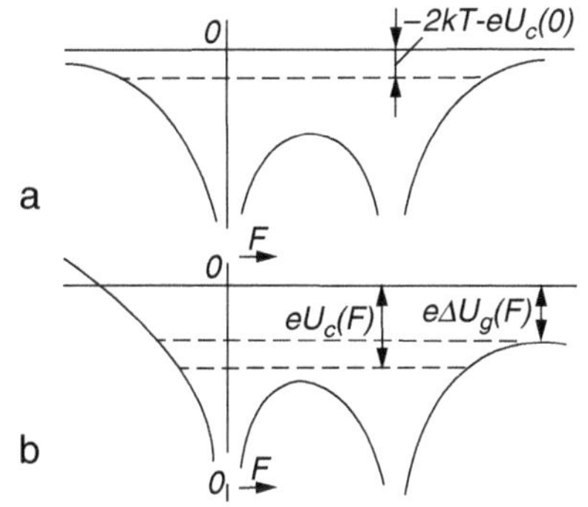

Fig. 2.17. (**a**) Coulomb funnel of an electron trap with an external electric field, indicating the lowering of the barrier for electron excitation (**b**) Two closer positioned Coulomb funnels with the same external field but a much lower reduction of the barrier between these centers

The change of the degree of negative differential conductivity becomes immediately visible when the $n_1(F)$ dependence is measured using the method of a pseudo-cathode described in Sect. 2.2.

It is evident from Fig. 2.18 that the slope of the negative differential conductivity (or electron density) is larger at lower doping which is directly shown in Fig. 2.18b where the quasi neutrality curve $n_2(F)$ is given for two otherwise similar CdS crystal platelets but with different copper doping. The corresponding conductivities are plotted in Fig. 2.18a.

The limited solubility of copper in CdS to 100 ppm and the substantial change of the photosensitivity prevents a broader range of doping for a meaningful investigation.

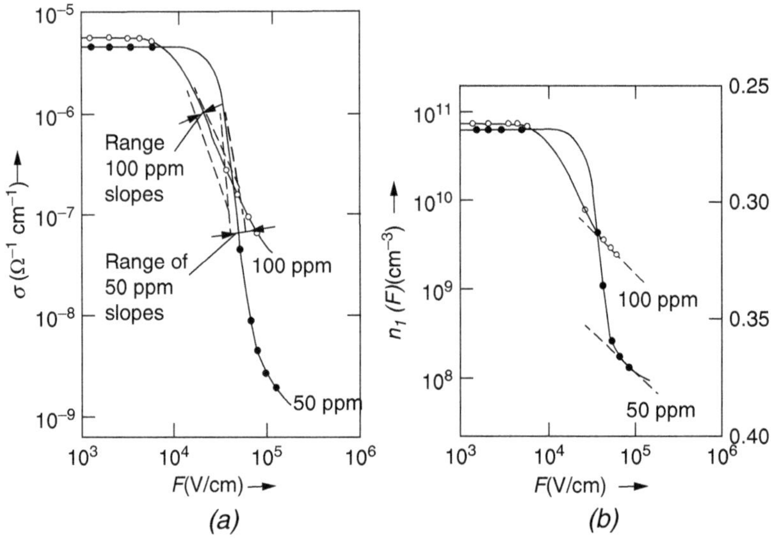

Fig. 2.18. Electron conductivity (**a**) and electron density (**b**) as well as the lowering caused by the Frenkel–Poole effect ($E_c - E_{Fn}$ (eV)) in a homogeneous CdS crystal as a function of the electric field, measured at 210 K with 1,013 photons/cm^2 at $\lambda =$ 515 nm, using high-field domain analysis. The doping density of copper is indicated in (**a**) (from Hadley et al. (1972)) measured for quasi neutrality using pseudo-cathode high-field domain analysis

2.2.2 Work Function Dependence on Photoconductivity

The high-field domain analysis can also be used for the investigation of the field dependence of the work function. Such a dependence was known for metal contacts at very high fields where tunneling through the barrier was assumed to be the reason for such dependence (i.e., lowering) of the work function.

However, when high-field domains occur, it is evident that the field near the metal/semiconductor interface cannot exceed the domain field which is usually <100 kV/cm, thereby excluding tunneling through the barrier. The carrier density in the domain is therefore a reasonable measure of the metal/semiconductor work function according to

$$n_{\mathrm{II}} \approx n_{\mathrm{c}} = N_{\mathrm{c}} \exp\left(-\frac{\psi^*_{\mathrm{MS}}}{kT}\right) \tag{2.9}$$

with ψ^*_{MS} the effective work function, reduced from the work function by the image potential

$$\psi^*_{\mathrm{MS}} = \psi_{\mathrm{MS}} - \sqrt{\frac{eF_{\mathrm{II}}}{4\pi\varepsilon\varepsilon_0}}, \tag{2.10}$$

with $n_{\rm II}$ and $F_{\rm II}$ the carrier density and electric field, respectively, at the second singular point (Fig. 2.4). The image force lowering with fields of $F_{\rm II} \simeq 10^5$ V/cm is on the order of 50 meV.

The current over the top of the barrier is carried by Richardson–Dushmann emission

$$j = en_{\rm c}v^*_{\rm rms} \simeq en_{\rm II}\mu F_{\rm II} \tag{2.11}$$

with $\mu F_{\rm II} \simeq v^*_{\rm rms}$ in the drift velocity saturation range this permits the identification of $n_{\rm c}$ with $n_{\rm II}$.

When high-field domains are generated in vacuum-cleaved copper-doped CdS crystals with immediately thereafter vapor-deposited electrodes, one can determine from the saturation currents in reverse bias the carrier density at the electrode, hence its work function. This yields effective work functions that are dependent on the optical excitation, i.e., on the carrier density $n_1(F)$ within the cathode-adjacent domain. This value is observed to vary substantially with optical excitation and is reduced to about one half of their values that are listed for vanishing currents which, for a gold electrode or other blocking metals, are on the order of 1 eV.

Figure 2.19 shows the linear dependence of domain width as a function of bias, and indicates a slight increase in domain fields with increasing optical generation rate (from 83 to 122 kV/cm for optical generation rates changing from 4×10^{14} to 5×10^{16} cm^{-3}/s, respectively). Since the carrier density within the domain also increases with increasing optical generation, one thereby observes a trend of a *decreasing work function with increasing optical excitation for all investigated contact metals.*

Measurements at temperatures between 150 and 300 K show also a substantial temperature dependence of the work function, and is given in Fig. 2.19c.

These findings seem to indicate that holes can significantly change the dipole layer at the metal/semiconductor interface. These holes are created by excitation and released by field quenching within the high-field domain directly adjacent to the interface, and thereby probably causing the corresponding lowering of the work function.

2.3 Summary and Emphasis

In semiconductors in which the conductivity decreases stronger than lin early with the electric field, certain inherent inhomogeneities must occur. These are high-field domains in an N-shaped negative differential conductivity regime and current channels when it is S-shaped. CdS is an ideal model substance that, because of its photoconductivity, can be shifted from one to the other regime. It is n-type and its behavior can easily be analyzed in a one-carrier model that for crystal platelets can be discussed in a one-dimensional transport and Poisson equation model.

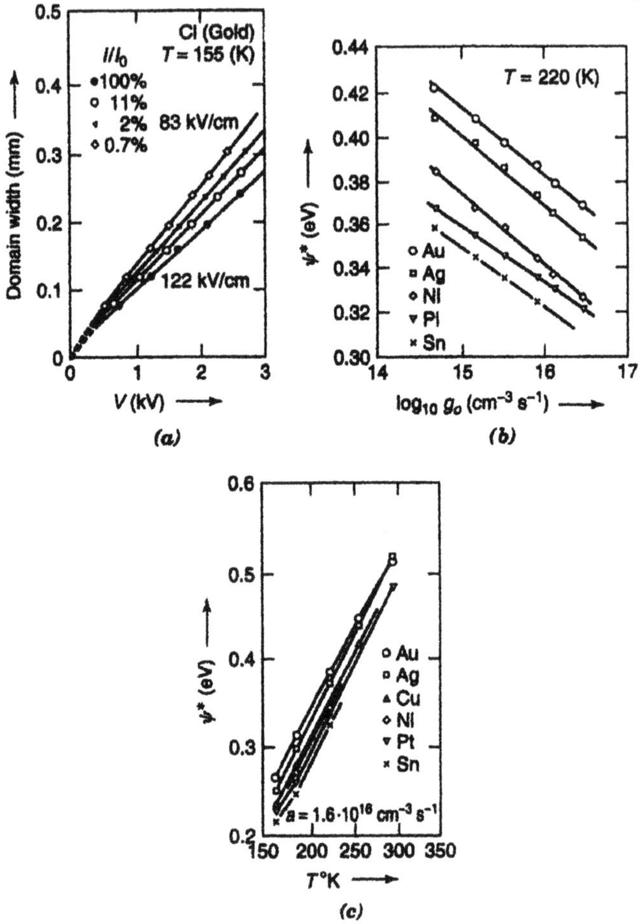

Fig. 2.19. (**a**) Domain width in CdS as a function of the applied voltage with light intensity as family parameter. Slopes yield the domain fields. (**b**) Effective work function as a function of the optical generation rate, and (**c**) as a function of the temperature for a number of blocking metal electrodes (After Stirn et al. 1971)

Such an analysis, as performed for stationary solutions, that is for stationary electrode attached domains, is highly transparent in a field-of-direction that identify the only possible types of solution curves, namely certain high-field domains.

Such stationary high-field domains, made directly visible by the Franz–Keldysh effect are extremely useful tools to directly and unambiguously determine the dependence of electron density and mobility on the acting electric field. They also permit to obtain direct evidence that the work function of metals to semiconductors depend on the photo-excited current through the CdS.

Such domains that are caused by field quenching, producing the decrease of carrier densities with increasing fields, are also a useful tool to obtain critical information about field quenching and its dependence on copper doping in CdS.

High-field domains can be used to unambiguously measure the field dependence of the electron density and mobility and its dependence on the doping density with Coulomb attractive copper centers in CdS. They also show that the metal/CdS work function depends on the optical excitation.

3

Moving High-Field Domains

Summary. High-field domains that remain attached to one electrode are stationary. Under a large variety of circumstances they separate from the electrode and move through the crystal. Such moving high-field domains can have significant technical importance, e.g., as the Gunn domain. Others can be helpful to identify parameter relations of a specific device, and still others may be used for demonstration or simply as estetical display. Unfortunately, some of these domain instabilities also produce disturbing electronic, mostly low frequency noise. In general, a solid understanding of the conditions that lead to such moving or unstable domains is essential for many devices that operate at high electric fields.

3.1 Stationary and Moving High-Field Domains

There are a number of conditions under which stationary domains attached to a cathode (or pseudo-cathode) as discussed in Chap. 2 are no longer sustained, but the domain separates itself from the cathode and starts moving through the crystal. A large variety of such domain motions and domain instabilities occur that can be followed directly, using the Franz Keldysh effect. To capture such kinetics, a film was produced that is available though the Springer Web site and could much more realistically represent the observations and compliment the description presented in this book. We reference specifically the film when such cinematographic presentation becomes more dramatic in the following sections.

One of the conditions that caused a moving high-field domain and observed in the very first experiment we performed, was given in Chap. 1 (see Fig. 1.3). This condition for separation of a domain from the cathode can be reached when the Schottky barrier is not strong enough (the contact is insufficiently blocking) or when the geometric configuration does not permit an expanding domain while attached to the cathode. The second condition was the reason for the domain separation in Fig. 1.3: here the expanding radius of the domain with sufficiently increased applied voltage caused an unstable situation in

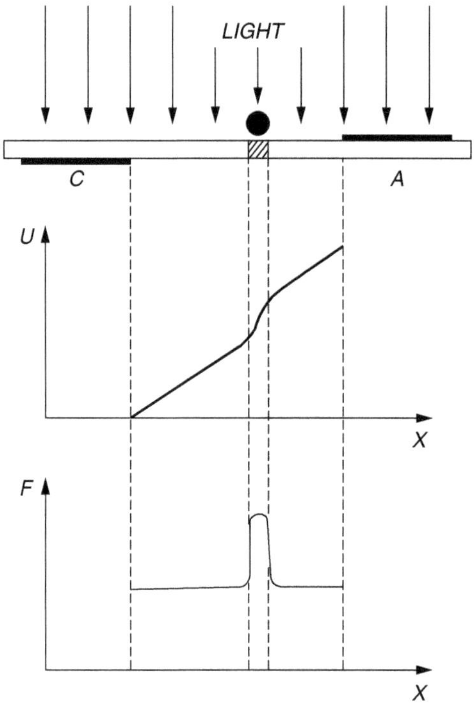

Fig. 3.1. Schematics of the pseudo-cathode method using a shadow across the crystal and the resulting potential and field distribution shown schematically below

which the minimum entropy production could only be maintained when a ring of approximately the same width separated from the cathode and expanded in radius, thereby maintaining essentially the same saturation current. A more extensive discussion of such moving domains will be given in later sections.

Here, we will mention an experimental method in which we can force a domain to remain free standing between the electrodes without being attached. For this, we use again the pseudo-cathode method that uses a shadow band across the crystal, separated from the cathode (see Fig. 3.1). If one removes the shadow fast and the applied voltage is large enough, the high-field domain remains near its initial position and then starts to move slowly toward the anode.

Such a motion can be understood when recognizing that at the cathode side of the domain the field increases steeply from the low-bulk value to the higher domain field, where the electron density consequently becomes much lower. At this edge, a thin space charge layer occurs due to the diffusion of electrons from the high-density region. This causes a negative space charge that maintains this edge (see Fig. 3.2). In the bulk, both densities of electrons and ionized donors compensated each other. On the anode side of the domain, a positive space charge occurs that is produced by the electron depletion

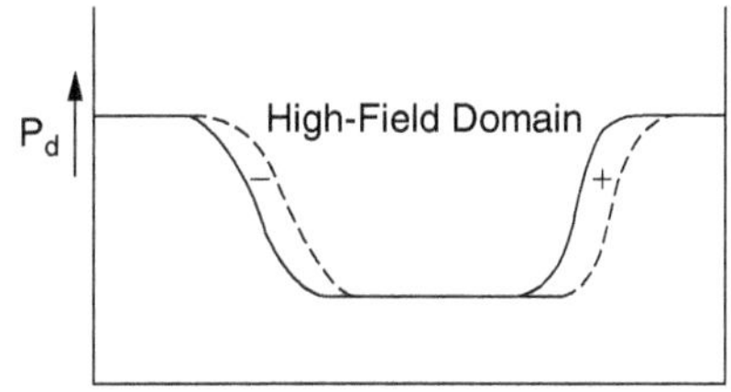

Fig. 3.2. Space charge regions at the edges of the free moving high-field domains caused by the diffusion of the mobile electrons, shown as *dashed curve* into the left side and out of the right side of the domain

from the edge. When considering the applied voltage as driving force and the highly mobile electrons moving faster toward the anode than the holes that are consecutively eliminated by recombination, one concludes that the domain must move toward the anode with a speed determined by reaction kinetics. This is at lower temperature a rather slow process, causing a rather slow motion of the domain (typically of 10^{-2} cm/s, as observed). However, as the re-emission of trapped electrons increases with increasing temperatures, one observes a speeding-up of the motion of the domains.

3.1.1 Self-Sustained Free Moving High-Field Domains

With the shadow, that initiates the high-field domain, the boundary condition that determines the electron density within the adjacent domain is also removed and is replaced by the earlier described and newly formed space-charge region at the left side of the domain. There is no calibration for the degree of field quenching that is given initially by the original domain. However, since the new space charge is self-created, it tends to increase the field and stops only when the domain region is fully quenched. One therefore expects and observes a slight further darkening of the domain (see Fig. 3.3) when it is separated from the shadow that typically causes a higher boundary electron density than the fully quenched case.

When the domain has time on its motion through the crystal to fully develop, one observes usually a smooth increase of the field from its cathode edge, followed by a decrease at the opposite edge, as shown in Fig. 3.4. However, this motion is quite sensitive to slight crystal defects, as can be observed best in the film and can cause some splitting or delay of parts of the moving domain as also shown further below.

With a rapidly removed shadow one has also the opportunity to initiate more than one high-field domain sequentially within a wide enough crystal, by starting another one with a shadow while the first one is already separated. In this fashion, as many as five high-field domains were created sequentially that all moved at low speed through the crystal toward the anode. The only condition one had to fulfill to maintain several domains, was to increase the

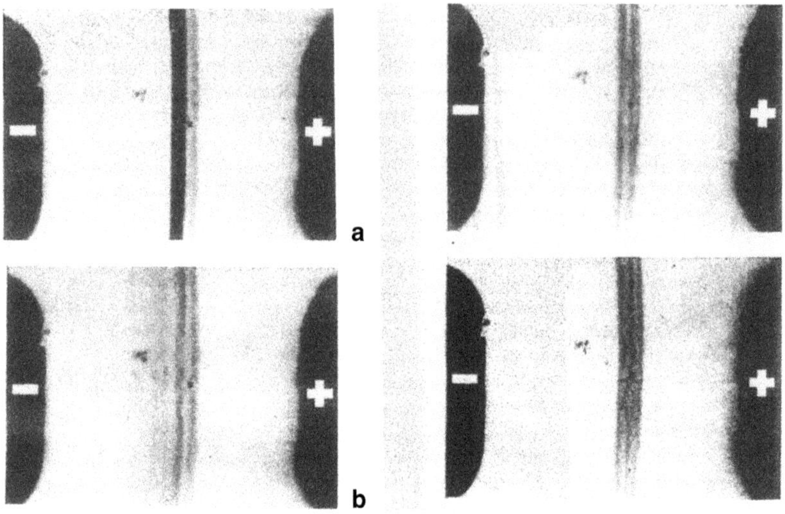

Fig. 3.3. (**a**) High-field domain extending beyond the right edge of a shadow. (**b**) The shadow is removed toward the left (blurred) leaving the domain self-sustained. (**c**) The domain darkens, indicating an increased domain field. (**d**) Further darkening of the center part of the domain and slight domain moving toward the anode (After Böer and Wilhelm 1963)

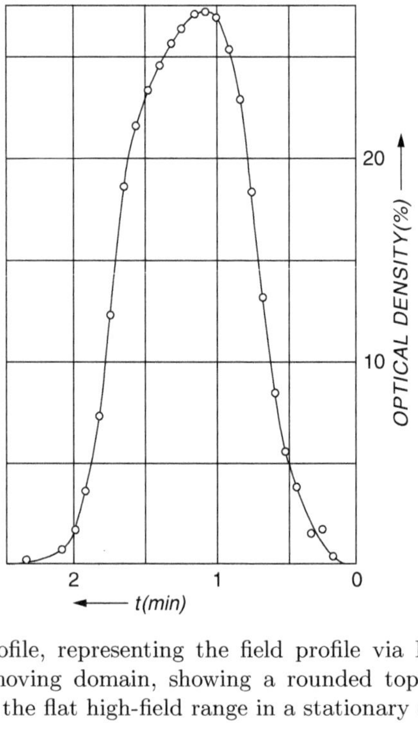

Fig. 3.4. Density profile, representing the field profile via Franz Keldysh effect darkening of a free moving domain, showing a rounded top of a thin high-field domain in contrast to the flat high-field range in a stationary domain

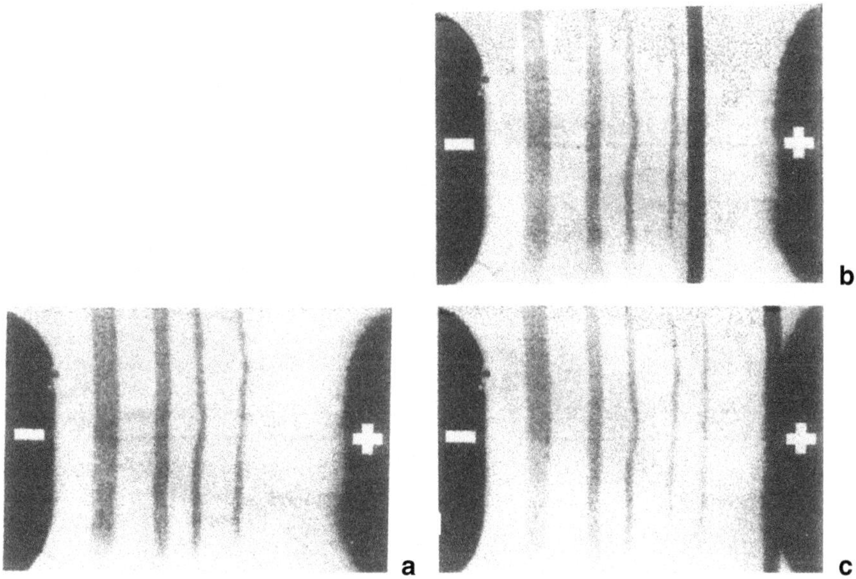

Fig. 3.5. (**a**) CdS crystal with four separate high-field domains in which in (**b**) a fifth one was initiated with a shadow, that was removed in (**c**), showing the fifth (though thin) high-field domain. The applied voltage was step-wise slightly increased with every additional domain (After Böer and Wilhelm 1964)

applied voltage with every additional domain so as to maintain the critical field for field-quenching within each domain. In the experimental results shown in Fig. 3.5, the applied voltage was not rapidly and sufficiently increased enough between the different domain initiations, causing the following ones to be thinner (and rapidly fading away if not sustained by higher applied voltages).

The width of the self-sustained domains increases linearly with the applied voltage, as long as these domains move essentially undeformed through the crystal as shown in Fig. 3.6. This is the same behavior as observed with stationary, cathode adjacent domains and is observed as long as the field quenching within the domain remains constant at its saturation range.

The domain motion is a good indication of the homogeneity of the crystal doping and the absence of substantial crystal defects (Fig. 3.7).

3.1.2 Unstable Moving High-Field Domains

At higher applied voltages, the high-field domain becomes unstable (see Fig. 3.8). First a small part of it may separate were some crystal defects prevent homogeneous increase of widths as shown in Fig. 3.8. This effect of partial separation is quite sensitive to small crystal defects or small inhomogeneities of doping and is clearly visible in Fig. 3.8e, f. In time, this separation

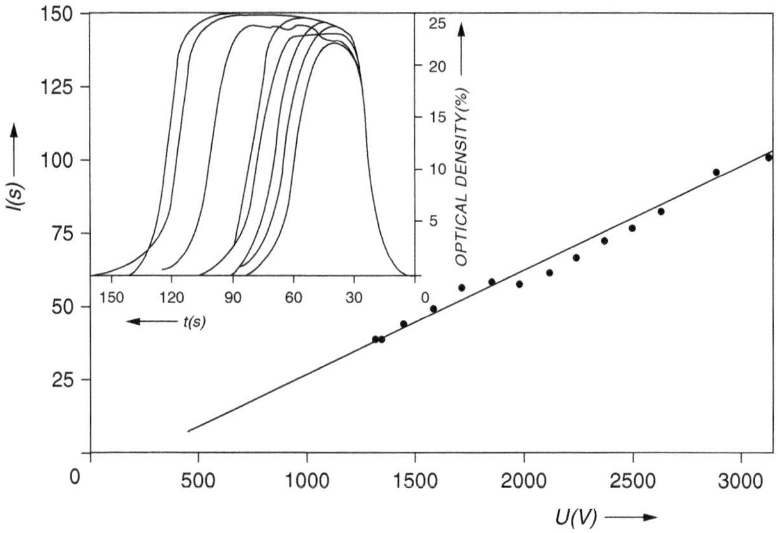

Fig. 3.6. The width of the self-sustained high-field domain increases linearly with applied voltage at the electrodes. The figure insert shows the corresponding density scan of eight domains as their projection passes over the pinhole in front of a photomultiplier (After Böer and Wilhelm 1964)

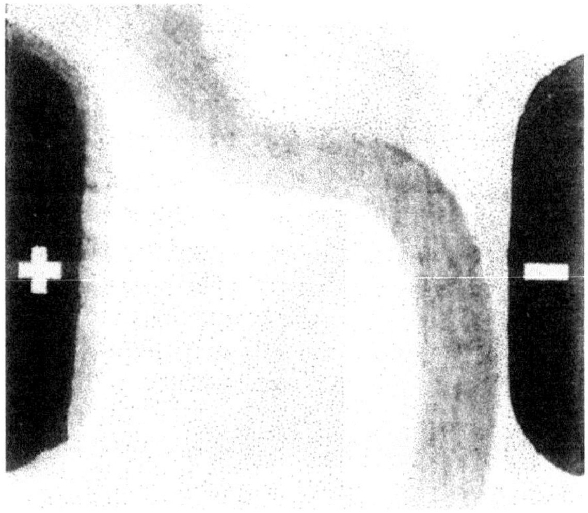

Fig. 3.7. Self-sustained high-field domain that moves toward the anode (observe the different polarity in this picture) as it was initiated by a pseudo electrode close to the cathode. The doping is inhomogeneous perpendicular to the applied voltage, resulting in a much faster motion of the domain at the top part of the crystal. Here it has almost reached the anode while near the bottom it has only slightly moved (After Böer and Wilhelm 1964)

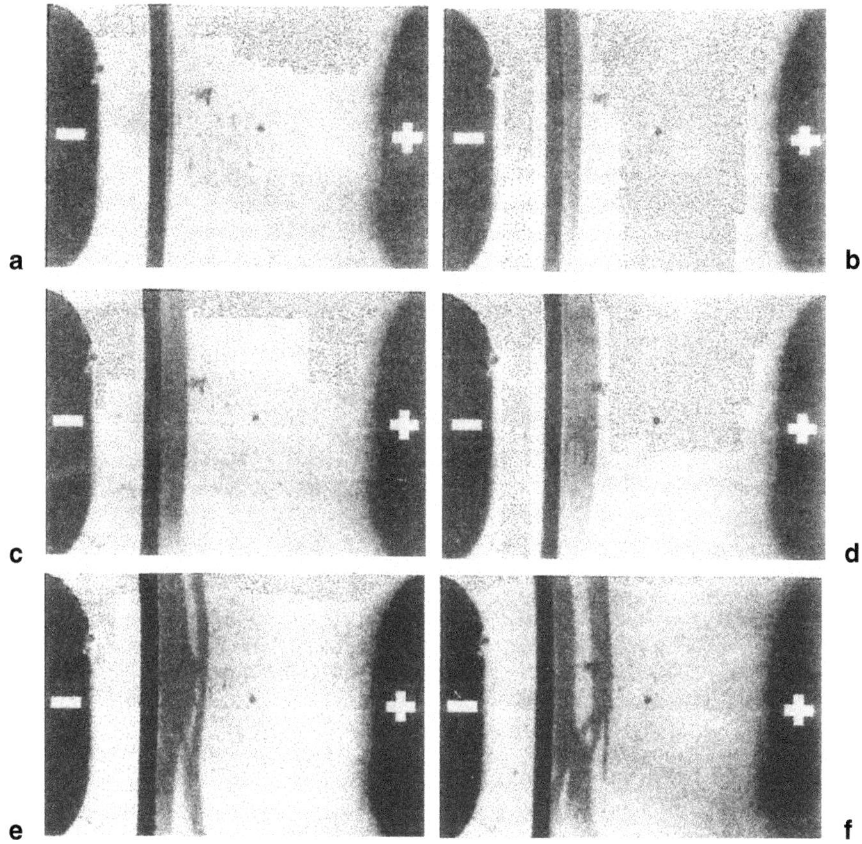

Fig. 3.8. Partial separation of the high-field domain with increasing applied voltage, hence increasing domain width, as a result of certain crystal defects (**a**–**e** for increased applied voltages of 1.0, 1.5, 2.0, 2.5, and 3.0 kV and kinetic of further separation at 3.0 kV shown in **f**) (After Böer and Wilhelm 1964)

increases and indicates the start of a domain instability. Again the film shows the kinetics more precisely.

Separating Domains as Sensitive Tools to Observe Crystal Inhomogeneities

In Fig. 3.9, a large variety of inhomogeneous deformation of the separating or fusing domains are shown that illustrate small deviations in the crystal doping that slows or accelerates their motion and enhances or hinders their separation. The changes of the domains in time can best be followed in the film. Here are four frames shown at the left that are 5 s separated and on the right the four frames are 10 s separated from each other. One also should recognize

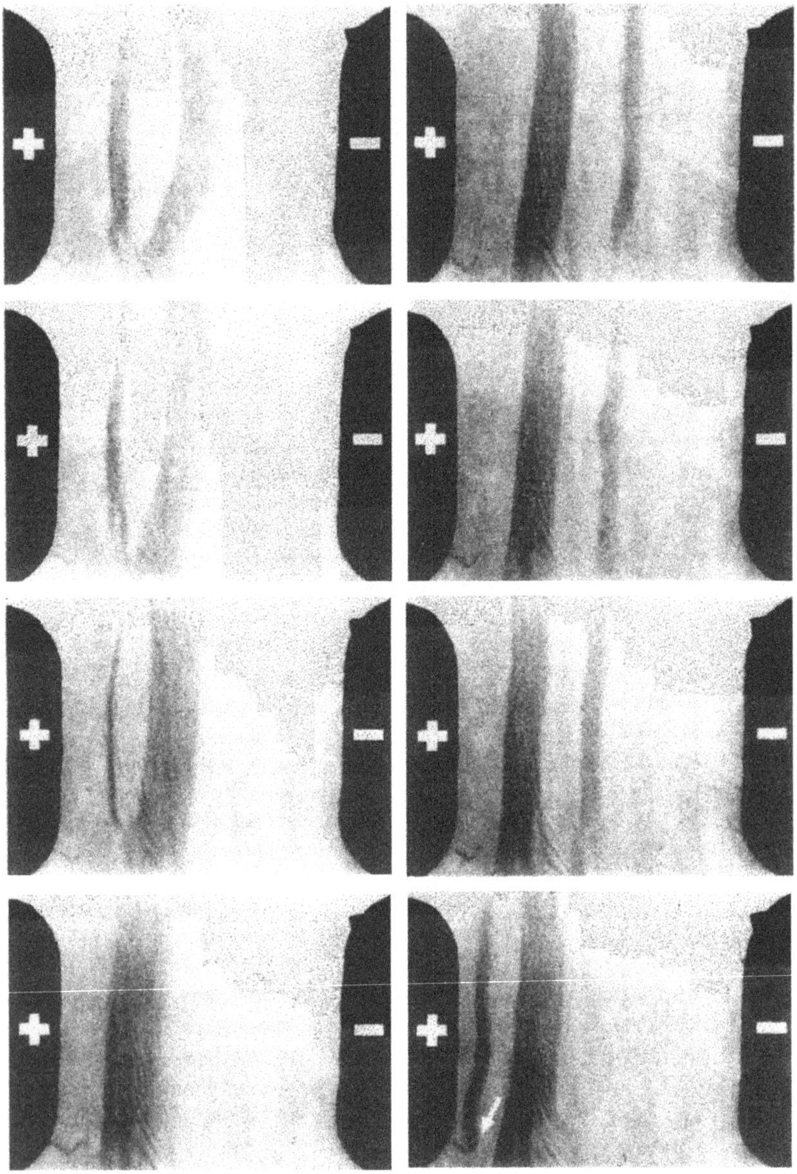

Fig. 3.9. Here, a sequence of sub-figures are given that show the deformation of domains occurring while they separate. In series (**a**)–(**d**) on the left side of the figure while they fuse together at 3,060 V and 1 μA and at the right series from (**a**) to (**d**) how they separate into two domains at 3,000 V and 1 μA. In (**d**), the arrow points at another defect that acts as an extension of the anode

that the CdS crystal was selected as one, that appeared very homogeneous with few striations and with attached, stationary domains that did not show marked inhomogeneities of any kind.

This is an indication that on the onset of instabilities, the moving domains are very sensitive to small deviations in doping that can locally influence the reaction kinetics including trapping and recombination. Thereby, it can sensitively change the condition for domain separation or the velocity of their motion. Also, crystal defects that are barely visible can have a marked influence on the kinetics.

All this may sound like academic interest; however, the performance of many devices depend on electronic kinetics and a number of yet unexplained deviation from perfect performance may be associated with field instabilities of a similar kind. These "deviations from perfect performance" are often collectively described as electronic noise. We will return to this more oscillatory behavior in Sect. 3.1.2.

Conditions for Breaking Up of Domains

In general, even without crystal defects, the attached domains become unstable and start to break up and move toward the anode. To show the dependence on the boundary condition, we have again used the pseudo-cathode concept by employing a shadow band, however, with various reduced illuminations within this band.

For further discussions we will now refer to the experimental current–voltage characteristic with an extrapolation to the nonmonotonic dashed branches. Here, the shadow pseudo-cathode is moved much closer to the anode so that the entire sample between shadow and anode can be filled with a high-field domain before instabilities occur (see below). Herewith $F = F_{\mathrm{II}}$ or $F{+}F_{\mathrm{III}}$ are obtained for the descending branch that is caused by field quenching, and the ascending branch caused by field excitation, respectively and are shown in curve o in Fig. 3.10).

With the shadow now moved closer to the cathode, one can expand the high-field domain within the crystal adjacent to the shadow. Different domains of various degree of field quenching can be generated depending on the light intensity within the shadow. With low-density gray filters for creating the shadow (curves 1 and 2 of Fig. 3.10), the current through the crystal is only little diminished and faint domains can be expanded almost reaching the anode. At much higher shadow densities (curves 3–8), the saturation branch is correspondingly lowered and with a higher applied voltage a range occurs in which the domains become unstable. This is indicated by the split of the characteristic (dashed part) that shows the range of domain splitting and current oscillations that occur concurrent with such instabilities.

Such moving domains and instabilities can be distinguished as three different kinds:

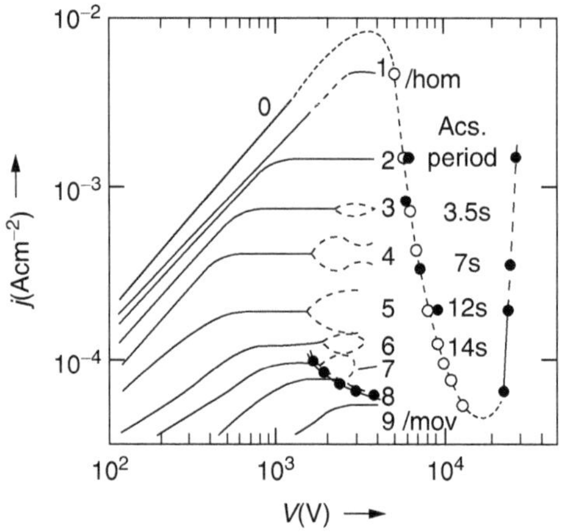

Fig. 3.10. Current–voltage characteristic of a CdS platelet with pseudo-cathode. *Dashed* are the branches due to field quenching and field excitation obtained from homogeneous filling of the region between pseudo-cathode and anode with an F_{II} domain, and from widening of an F_{III} domain, respectively. Family parameter of all other characteristics is the effective boundary concentration n_c at the edge of the pseudo-cathode that is stepwise reduced in light intensity from *curves* 1 to 9. For more, see text (After Böer and Voss 1968)

(1) Unstable domains that appear with an otherwise observed stationary cathode-adjacent domain at its anode-adjacent end and separated from this domain, moving toward the anode while disappearing
   (a) Before they reach the edge of the previous domain or
   (b) After passing through the edge of the original domain and
(2) Undeformed domains moving from cathode to anode

The ranges of the different types of instabilities in the current voltage characteristic are identified in the graph shown in Fig. 3.11.

When moving domains of type 1a or 1b appear, the current oscillates around its stationary saturation value, exceeding this value when it reaches the anode side limit and its lower value when it started again near the cathode side, as indicated by the spread of the dashed curves in Fig. 3.10.

However, in most cases, at the beginning of the instability the kinetic appears to be more spontaneous, starting at different points of the domain, other appearing as "bubbling" rather than wave motion (best seen in the film); this is often also described as a chaotic motion, even though in its microscopic detail, sub-parts of this instability appear periodic. When observed in an oscilloscope, this onset appears simply as electronic noise.

After the separated domain now becomes organized (in time and at slightly higher applied voltages), it starts moving through the crystal. Moving high-field domains have a much larger field as discussed in Sect. 3.1.2 and maintain

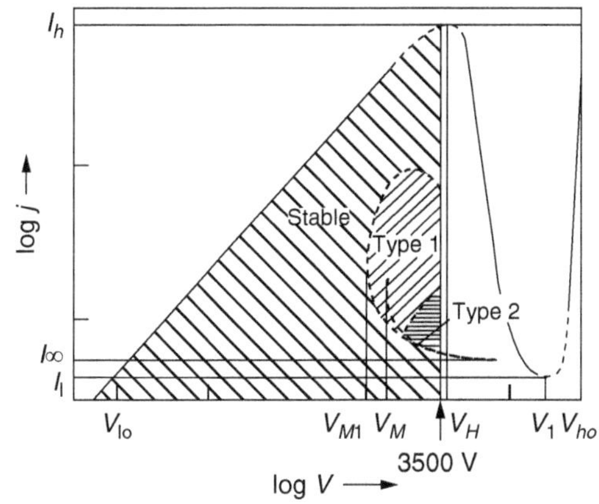

Fig. 3.11. Current–voltage ranges in which one obtains stationary cathode-adjacent high-field domains, unstable moving domains of type 1a or 1b (see text) and undeformed moving domains of type 2. The current–voltage characteristic between V_{M} and V_{H} is measured while a high-field domain moves through the crystal, but is well separated from the electrodes. The minimum currents are identified by j_l and j_∞ for the moving domains and the minimum current extrapolated from the singular point analysis

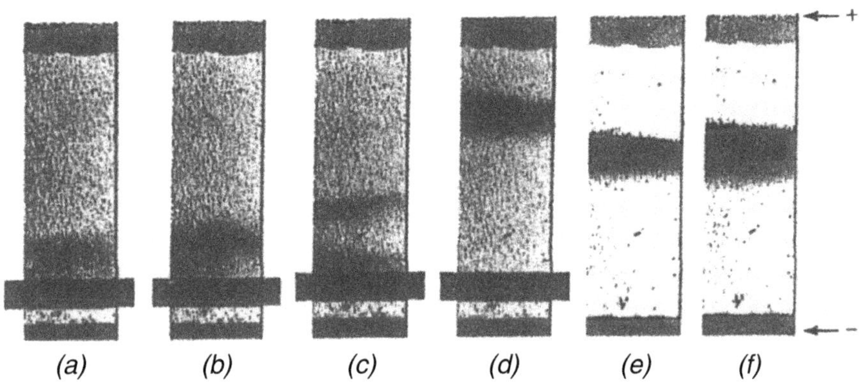

Fig. 3.12. Photographs of a CdS platelet with pseudo-cathodes (**a**–**d**) and without such pseudo-cathode at higher applied voltages (**e** and **f**) measured at 200 K and 508.5 nm illumination. (**a**) stationary domain at 1,650 V applied, (**b** and **c**) unstable domain of type 1a at 2,000 V (**d**) type 1b unstable domain at 2,150 V, (**e**) type 2 moving domain (shadow removed) at 2,400 V, and (**f**) same as (**e**) but at 4,000 V

a much lower saturation current as indicated by j_{mov} that lies close to the minimum of curve (o) in Fig. 3.10.

Examples for these three kinds of domains are given in Fig. 3.12. The degree of darkening through the shift of the absorption edge from the Franz–Keldysh effect is an indication of the acting field strength within the domain. It

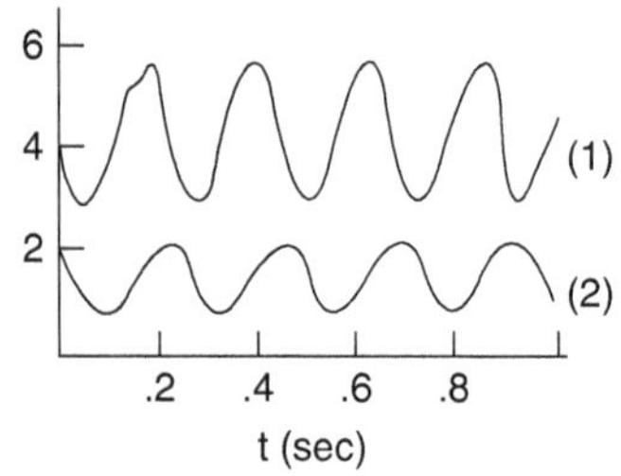

Fig. 3.13. Double trace oscilloscope picture of the current (1) through the crystal and the optical absorption (2) in the middle of the high-field domain measured as projected through a pinhole by a photomultiplier and caused by Franz–Keldysh shift of the absorption edge (After Böer and Ward 1967)

clearly demonstrates the much larger domain field in the free moving domain as explained in Sect. 3.1.2. Concurrent with these moving domains are well-defined current oscillations.

Moving Sub-Domains

A totally different type of moving domain is observed that moves at faster speed and is observed under certain doping conditions *within* a slow moving high-field domain. These sub-domains move in the opposite direction with a considerably higher domain velocity (by a factor of ≈100). The motion of such sub-domains, when organized, is also accompanied with current oscillations, as shown in Fig. 3.13 The current minimum is observed while the sub-domain is in the center of the high-field domain, i.e., well separated from the domain edges.

These sub-domains are probably due to instabilities in the hole assembly within the highly quenched main domain region. Here, such sub-domains may be used as indication that the conductivity regime is inverted from the n-type in normal CdS to p-type within the quenched sub-domain (Böer and Ward 1967).

Again, in many cases, as shown in the film, the sub-domains may not be fully organized, but move often out of phase in different parts of the main domain.

3.2 Stability Criteria and Moving High-Field Domains

To obtain theoretical criteria under which conditions high-field domains are stable and when they become unstable, we need to include into the discussion of the transport and Poisson equation the time dependence:

$$j = en\mu_{\mathrm{n}}F + \mu_{\mathrm{n}}kT\frac{\mathrm{d}n}{\mathrm{d}x} + \varepsilon\varepsilon_{\mathrm{o}}\frac{\partial F}{\partial t} \tag{3.1}$$

and

$$\frac{\partial F}{\partial x} = \frac{e}{\varepsilon\varepsilon_o}(n + n_t - p_t). \tag{3.2}$$

This, in principle adds one more dimension and makes a simple two-dimensional field-of-direction analysis impossible. However, since at least one solution appears as a propagation of an undeformed domain with constant speed of propagation,[1] one can discuss this type of domain by a simple coordinate transformation:

$$x' = x - ct, \tag{3.3}$$

yielding a modified two-dimensional representation for the transport equation:

$$j = en\mu_n F + \mu_n kT\frac{\mathrm{d}n}{\mathrm{d}x} - \varepsilon\varepsilon_o\frac{\mathrm{d}F}{\mathrm{d}x'} \tag{3.4}$$

and the transformed Poisson equation

$$\frac{\mathrm{d}F}{\mathrm{d}x'} = \frac{e}{\varepsilon\varepsilon_o}(n + n_t - p_t), \tag{3.5}$$

which again is an autonomous system and permits projection into the n, F plane and any x', i.e., it permits a field-of-direction analysis for the undeformed moving solution (neglecting the electrode effects).[2] When introducing the kinetics, that means to investigate the redistribution of carriers over traps and bands due to excitation, recombination and trapping. This requires for a proper analysis that we introduce a reaction kinetic model along the lines already indicated in one of the previous sections. For the ease of understanding, we will introduce these equations again here and introduce the appropriate parts into the transport equations in Sect. 3.2.1.

3.2.1 Trap-Controlled Kinetics

When the negative differential conductivity is caused by a decrease of carrier density with field, a redistribution of carriers over traps takes place and slows the motion of the domains. This can be followed through the reaction kinetic

[1] This means that any solution $\phi(x,t)$ can be written as $\phi(x - ct)$, where ϕ stands for any variable n, n_t, p_t, F, and j, and c is the constant speed of the moving domain. This, together with (3.3) leads to $\frac{\partial}{\partial t} = -c\frac{\partial}{\partial x} = -c\frac{\mathrm{d}}{\mathrm{d}x}$.

[2] Having an advantage of easy interpretation, the field-of-direction analysis, however, implies that $\mathrm{d}j/\mathrm{d}x \equiv 0$, that is that the current does not change in time. This analysis therefore does not describe current oscillations that occur when the domains approach the electrodes and are annihilated there, or are created when emerging from the cathode.

model. A minimum model capable of the discussion of negative differential conductivity contains only one type of electron traps (a deep donor) and one acceptor for partial compensation. Reaction kinetics enters via the change of space charges near the edges of the domain when the domain propagates. The space charge

$$\varrho = e(n + n_{\mathrm{a}} - p_{\mathrm{d}}) \tag{3.6}$$

with $n_{\mathrm{a}} \approx N_{\mathrm{a}}$, and $p_{\mathrm{d}}(t) = N_{\mathrm{d}} - n_{\mathrm{d}}(t)$ is determined by

$$\frac{\mathrm{d}n_{\mathrm{d}}}{\mathrm{d}t} - e_{\mathrm{tc}} n_{\mathrm{d}} - e_{\mathrm{ct}} n (N_{\mathrm{d}} - n_{\mathrm{d}}). \tag{3.7}$$

With domains moving slowly enough, so that the time it takes for the domain to move across a Debye length is larger than the relaxation time for electrons in donors, one can assume a quasi-steady state, that is $\mathrm{d}n_{\mathrm{d}}/\mathrm{d}t \equiv 0$, and has

$$n_{\mathrm{d}} = \frac{c_{\mathrm{ct}} N_{\mathrm{d}} n}{e_{\mathrm{tc}} + c_{\mathrm{ct}} n}. \tag{3.8}$$

From (3.2), (3.5), and (3.8) one now obtains a determining set of equations for the field-of-direction analysis:

$$\frac{\mathrm{d}n}{\mathrm{d}x'} = \frac{e}{\mu_{\mathrm{n}} kT} \left[n\mu_{\mathrm{n}} F - \frac{j}{e} - c \left\{ n + \frac{c_{\mathrm{ct}} N_{\mathrm{d}}}{e_{\mathrm{tc}} + c_{\mathrm{ct}} n} n - (N_{\mathrm{d}} - N_{\mathrm{a}}) \right\} \right] \tag{3.9}$$

and

$$\frac{\mathrm{d}F}{\mathrm{d}x'} = \frac{e}{\varepsilon\varepsilon_{\mathrm{o}}} \left[n + \frac{c_{\mathrm{ct}} N_{\mathrm{d}}}{e_{\mathrm{tc}} + c_{\mathrm{ct}} n} n - (N_{\mathrm{d}} - N_{\mathrm{a}}) \right]. \tag{3.10}$$

We will now again use the auxiliary functions $n_1(F)$ for which the space charge vanishes (the neutrality curve), and $n_2(F)$ for which the diffusion current vanishes (the drift current curve), however now within the moving x' system.

Singular Point Analysis

For the analysis of the type of solution curves we have successfully employed the field-of-direction analysis within the n, F projection. This analysis involves specifically the three singular points of $n(F)$. Now, in the kinetic system (x') the singular points lie at the same position as in the stationary system (x). For the type of singular points,[3] however, such knowledge is essential for classifying the types of possible solutions (Böer and Dussel 1967).

[3] Such as a node, spiral point or vortex for the singular point II, and always a saddle point for the singular points I and III, as shown later.

A singular points evaluation requires a linear expansion close to the singular point $(n_{\mathrm{i}}, F_{\mathrm{i}})$:

$$n = n_{\mathrm{i}}(1 + \eta) \tag{3.11}$$

and

$$F = F_1(1 + \zeta) \tag{3.12}$$

with slopes in the vicinity of $(n_{\mathrm{i}}, F_{\mathrm{i}})$ given by

$$\frac{\mathrm{d}\eta}{\mathrm{d}x'} = A\eta + B\zeta \tag{3.13}$$

and

$$\frac{\mathrm{d}\zeta}{\mathrm{d}x'} = C\eta + D\zeta. \tag{3.14}$$

The solution of this system can be written as

$$\eta = \xi_1 \exp(\lambda^+ x') + \xi_2 \exp(\lambda^- x') \tag{3.15}$$

and

$$\zeta = \xi_3 \exp(\lambda^+ x') + \xi_4 \exp(\lambda^- x') \tag{3.16}$$

with

$$\lambda^{\pm} = \frac{A + D}{2} \pm \sqrt{\frac{(A + D)^2 + 4(BC - AD)}{4}} \tag{3.17}$$

and indicates for the discriminant

$$BC - AD \leq 0$$

a saddle point and

$$BC - AD < 0$$

a node, spiral, or vortex, depending on the propagation velocity of the domain, as shown in Fig. 3.14.

The field of direction in the vicinity of two types of singular points II is given in Fig. 3.15, where (a) is a stable and (b) is an unstable node (from the latter all directions of possible solutions point away from the node).

The characteristic velocities indicated in Fig. 3.14 are given by

$$v_{\mathrm{o}} = \left| \frac{v - \dfrac{\mu_{\mathrm{n}} kT}{e} \cdot \dfrac{e_{\mathrm{tc}} \tau_{\mathrm{g}}^2}{v \tau_{\mathrm{r}} \tau_{\mathrm{kr}}}}{1 + \dfrac{e_{\mathrm{tc}} \tau_{\mathrm{g}}^2}{\tau_{\mathrm{r}}}} \right|_{n_{\mathrm{II}}, F_{\mathrm{II}}} \quad ; \quad v = \mu_{\mathrm{n}} F \tag{3.18}$$

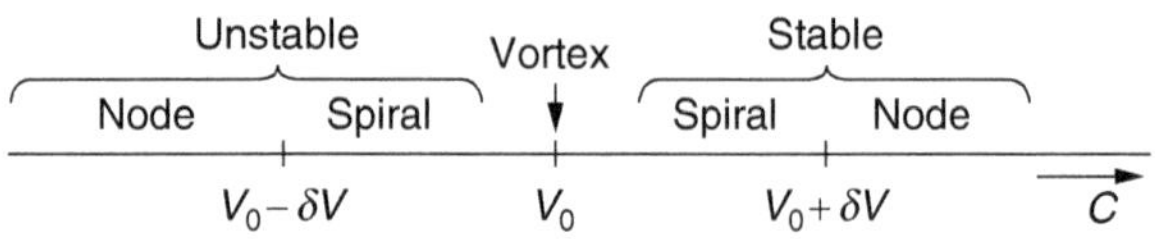

Fig. 3.14. Behavior of the singular point II of the system of transport equations (3.9) and (3.10) as a function of the domain velocity c

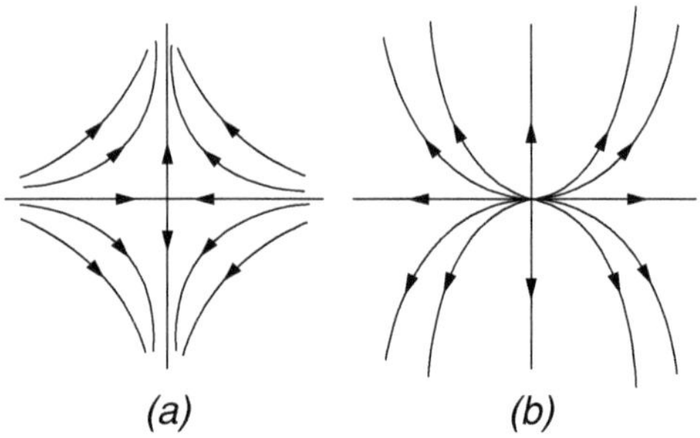

Fig. 3.15. Field of direction in the vicinity of (**a**) a saddle point, and (**b**) an unstable node

and

$$\delta v = \left| 2\sqrt{-\frac{\mu_{\mathrm{n}} kT}{e}\left(\frac{1}{1+\dfrac{e_{\mathrm{tc}}\tau_{\mathrm{g}}^2}{\tau_{\mathrm{r}}}}\right)^2\left(\frac{1}{\tau_{\mathrm{kd}}}+\frac{e_{\mathrm{tc}}\tau_{\mathrm{g}}^2}{\tau_{\mathrm{kd}}}\left[\frac{1}{\tau_{\mathrm{kd}}+\dfrac{1}{\tau_{\mathrm{kr}}}}\right]\right)}\right|_{n_{\mathrm{II}},F_{\mathrm{II}}} \tag{3.19}$$

with the characteristic time constants defined as

$$\tau_{\mathrm{g}} = \frac{1}{e_{\mathrm{tc}} + c_{\mathrm{ct}} n}; \quad \text{and} \quad \tau_{\mathrm{r}} = \frac{1}{c_{\mathrm{ct}} N_{\mathrm{d}}} \tag{3.20}$$

and of

$$\tau_{\mathrm{kd}} = \frac{1}{\dfrac{e}{\varepsilon\varepsilon_{\mathrm{o}}} n \dfrac{\partial v}{\partial F}}; \quad \text{and} \quad \tau_{\mathrm{kr}} = \frac{1}{\dfrac{e}{\varepsilon\varepsilon_{\mathrm{o}}}} n\mu_{\mathrm{n}} \frac{\partial \ln c_{\mathrm{ct}}}{\partial \ln F} \tag{3.21}$$

for the given reaction kinetic model.

From this analysis one sees that undeformed moving domains can only exist at the vortex, that is, here the domain velocity is equal to v_{o}. Now, to evaluate moving domains we must go beyond the proximity of the singular point II. However points I and III are of no interest here since they are saddle points and cannot support periodic solutions. In Sect. 3.2.1, we will discuss the solution of the transport and Poisson equations further away from the singular point II.

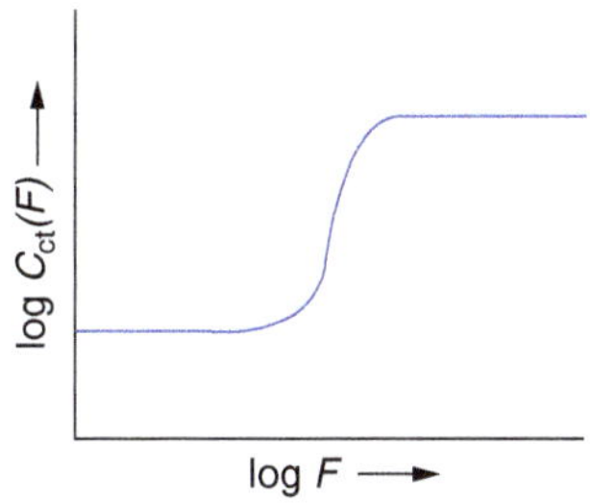

Fig. 3.16. Schematics of the field dependent recombination coefficient $c_{ct}(F)$ as assumed in the following discussions of the $n_1(F)$ and $n_2(F)$ curves that simulates field quenching in a simplified reaction kinetic model

Field-of-Direction Analysis for Kinetic Solutions

To quantitatively analyze the field of directions, we will now introduce a field-dependent recombination coefficient $c_{ct}(F)$ to simulate in a simplified model the field quenching without carrying additional adjustable parameters. With such field dependence of the recombination coefficient, as shown schematically in Fig. 3.16, we obtain the $n_1(F)$ curves shown in all previous field-of-direction figures.

The diffusion neutrality curve $n_2(F)$, however, is now also modified and depends sensitively on the domain velocity. It is given by the roots of the quadratic equation

$$n_2^2 + n_2\left[\frac{e_{\mathrm{tc}}}{c_{\mathrm{ct}}} - \frac{j + ecN_{\mathrm{a}}}{c_{\mathrm{ct}} e\mu_{\mathrm{n}} F}\right] + \frac{e_{\mathrm{tc}}}{c_{\mathrm{ct}} e\mu_{\mathrm{n}} F}\left[ec_{\mathrm{ct}}(N_{\mathrm{d}} - N_{\mathrm{A}}) - j\right] = 0 \qquad (3.22)$$

and is shown in Fig. 3.17 for increasing domain velocities from panel (a) to (e). Only panel (a) has the form similar to the one shown in all the previous chapters (except from the slight deviation from the linear behavior of $n_2(F)$ shown in this panel. The substantial changes shown in the following panels depend on the relation of the domain velocity c with the critical velocity

$$c_{\mathrm{o}} = \frac{j}{e(N_{\mathrm{d}} - N_{\mathrm{a}})} \simeq \left.\frac{v\tau_{\mathrm{r}}}{\tau_{\mathrm{g}}}\right|_{E_{\mathrm{II}}}, \qquad (3.23)$$

as indicated in the caption of Fig. 3.17. The completely different shapes of the $n_2(F)$ curves in panels (b)–(e) deserve a closer inspection. Closed solutions must lie between the singular points I and III (Poincaré's theorem) and circle the point II. Adding this to circling II at $c = v_{\mathrm{o}}$ as shown in panel (c), there are also solutions that are spiraling toward II for $c > c_{\mathrm{o}}$ as shown in panel (d) and discussed in the following section. However, when entered from the outside, the stable vortex point II for $c = v_{\mathrm{o}}$ yields only unstable solutions. Therefore stable, periodic solutions are only possible for domain velocities between c_{o} and v_{o}:

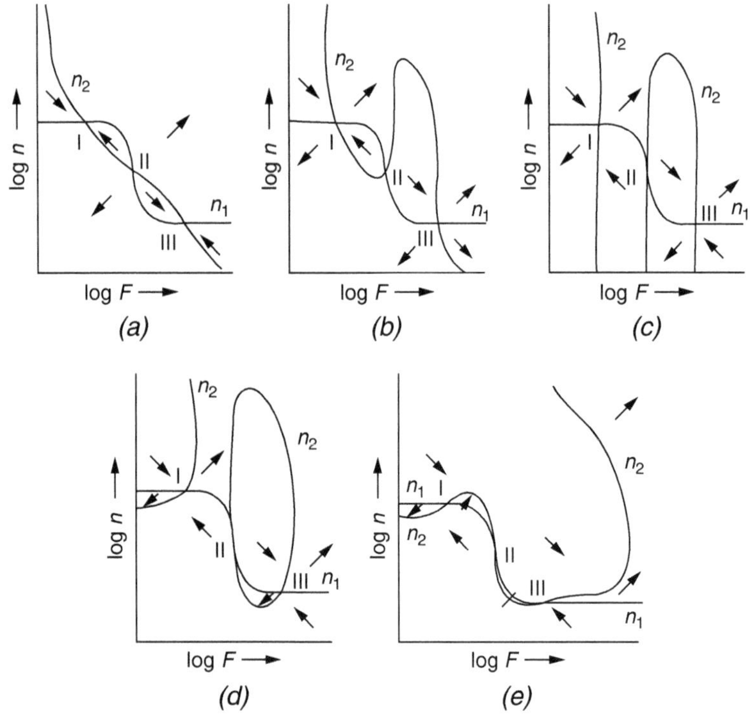

Fig. 3.17. Field-of-direction with neutrality curve $n_1(F)$ and diffusion neutrality curve $n_2(F)$ for undeformed moving high-field domains and different domain velocities in the panels (**a**)–(**e**) but constant currents. (**a**) for $c < c_\mathrm{o}$ and $c < j/(eN_\mathrm{a})$; (**b**) for $c < c_\mathrm{o}$ but $c > j/(eN_\mathrm{a})$; (**c**) for $c = c_\mathrm{o}$; (**d**) for $c > c_\mathrm{o}$; and (**e**) for $c \gg c_\mathrm{o}$. The critical velocity c_o is given in (3.23)

$$c_\mathrm{o} = \left(v\frac{\tau_\mathrm{r}}{\tau_\mathrm{g}}\right)\Big|_{E\mathrm{II}} < c < c_\mathrm{o} + \left[\frac{\mu_\mathrm{n} kT}{ev\tau_\mathrm{kr}}\right]\Big|_{N_\mathrm{II},E_\mathrm{II}} . \tag{3.24}$$

We will discuss this in more detail in the following section.

Periodic Solutions for Trap-Controlled Domains

Periodically moving, high-field domains require circling the singular point II, and remaining separate from the other two singular points as shown in panels (a) and (b) of Fig. 3.18. In principle, there is a large variety of such closed solution circling II possible, that are closer, and further separated from the singular point II.

As discussed in Sect. 3.2, the free moving domain is removed from the boundary condition near the singular point I and is self-sustained by the space charge layers at their edges, causing the field to increase close to carrier depletion within the domain that is also closer to the singular point III. This

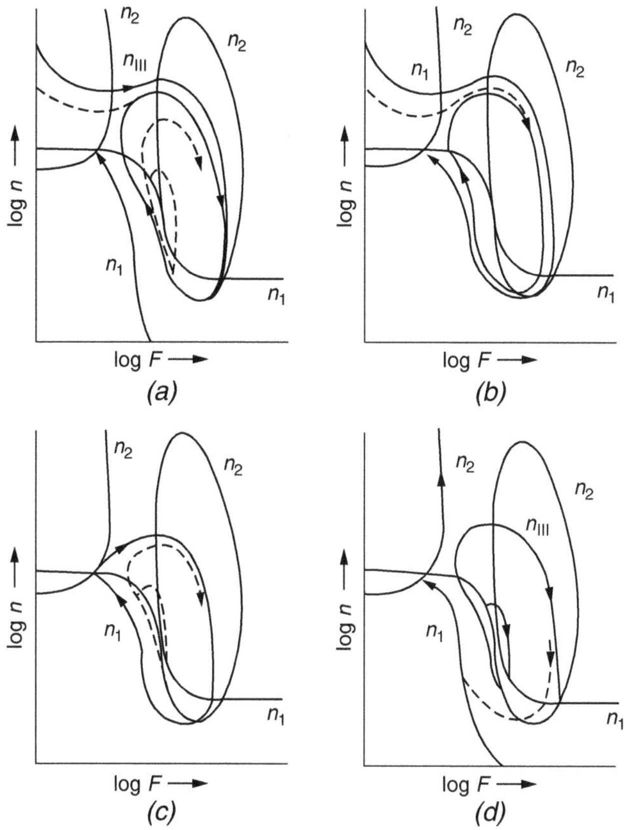

Fig. 3.18. Field-of-direction diagram as in Fig. 3.17 with possible solution curves that reach the saddle points I and III in panels (**c**) and (**d**). *Heavy arrows* in (**a**) and (**b**) indicated possible closed solutions. All solutions are assumed to start from *dashed arrows*

is also in agreement with the experimental observation shown in Fig. 3.12d–f. The current also tends to remain close to the minimum entropy production, indicated as the lower dashed line in Fig. 3.19b. None of the periodic solutions can come close to the singular point III, hence none of the moving high-field domains can show a completely flat top but rather curved maximum field within the domain as a closer observation of such moving high-field domain photographs using the Franz–Keldysh effect shows.

Consequently, only stationary high-field domains can have a flat field top region that was used in earlier chapters to unambiguously determine field dependent parameters.

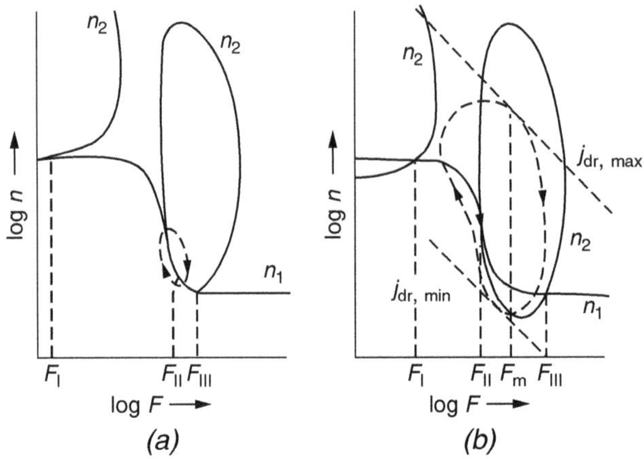

Fig. 3.19. Field-of-direction diagram as in Fig. 3.18a with possible solution curves shown by *closed curve dashed arrows*. Two alternatives are shown (**a**) with low amplitude and (**b**) with high amplitude, the latter extending between the minimum and maximum entropy-related currents identified by *dashed lines* with $j_{\mathrm{dr,min}}$ and $j_{\mathrm{dr,max}}$, respectively

3.3 Summary and Emphasis

Whenever the conductivity decreases more than linearly with increasing electric field, a new type of solution of transport- and Poisson-equation is attained which divides the semiconductor into two parts, one with a high field, the high-field domain, and the other part with a low field. This field distribution can be stationary or nonstationary. In stationarity, the field is essentially constant in each of these regions. Both regions are joined by a narrow space charge region ("step") of only a few Debye lengths width.

When the high-field domain moves from one to the other electrode, it disappears when it reaches the electrode, while a new domain evolves from the first electrode and the process repeats itself periodically. Stationary domains are usually observed when the negative differential conductivity is caused by a field-dependent decrease of the carrier density, while no stationary solution can be maintained when the change in conductivity is caused by a mobility that decreases more than linearly with increasing field (Gunn effect).

Other examples of stationary or moving domains are provided by resonant tunneling in super lattices, real-space transfer in modulation-doped heterostructures, or nonlinear generation–recombination kinetics in p–Ge or semi-insulating GaAs (see Chapter 5).

With increasing bias, stationary high-field domains expand at the expense of the low-field region, until the entire semiconductor is filled by the high-field domain. With further increased bias, a second type of domain with a substantially higher field develops from the opposite electrode and expands as required for absorbing the entire applied voltage.

A relatively simple field-of-direction analysis of the family of solution curves of Poisson- and transport-equations permits us to obtain a general overview of the different types of domains and to deduce unambiguous information on the field dependence of carrier densities and mobilities, as well as on the carrier density and field at the metal/semiconductor boundary.

Stationary high-field domains can be used to measure the work-function of a metal/semiconductor contact and its dependence on various parameters, providing direct evidence of a reduction of this work function with increasing optical generation rates in photoconductors.

When Hall electrodes are located within a stationary high-field domain that is exposed to a magnetic field, the dependence of the electron mobility as a function of the domain field can be determined unambiguously.

The use of a pseudo-electrode by a band of reduced illumination in photoconductors permits the direct measurement of the field-dependent carrier density in semiconductors.

Moving high-field domains are slow when related to carrier redistribution over traps and can be very fast when trapping can be neglected in domains caused by a field-induced decrease of carrier mobilities. The Gunn effect is an example where such steeper-than-linear decrease of the mobility with increasing field in GaAs is used for high frequency ac generation.

High-field domains can be visually observed by the shift of the optical absorption edge due to the Franz–Keldysh effect and its kinetics can be studied easily.

The kinetic behavior of unstable domains can also be used to explain a wide variety of electronic noise especially in the low-frequency range.

The kinetic effects of such moving or unstable high-field domains can best be recorded cinematographically in a film that is made available to the readers of this book through a Web site of Springer Verlag, Berlin. This film contains a wealth of the experimentally observed evidence in such moving domains, their generation and their dependence on numerous crystal parameters. The film also explains in graphics the related theories and application.

High-field domains are a valuable tool to unambiguously investigate the field-dependence of important parameters in semiconductors. They are also cause for current oscillation, and when chaotic, the cause for low frequency electronic noise.

4

Negative Differential Conductivity Caused by Mobility

Summary. In certain semiconductors the electron mobility decreases strongly with increasing field. A typical example is GaAs in which electrons in the conduction band can be heated by the field and then start to populate a higher sub-band with lower effective mass. This process is fast and not hindered by redistribution of electrons over traps and recombination centers. As a consequence, high-field domains are created and the current oscillates when these domains move between the electrodes and temporarily disappear when they reach an electrode. These are high-frequency oscillations and are of technical interest for such current generation (Gunn-effect).

4.1 Gunn Domains

High-field domains can be generated only when a second singular point occurs in the field of directions. This requires a range of negative differential conductivity that can also be obtained by a steeper than linear decrease of the mobility with increasing field, and was first predicted by Böer (1959).

Such a decrease in mobility can occur by the repopulation of electrons over bands with substantially different effective masses, as observed in GaAs with an effective mass of 0.07 m_0 in the lowest (central) valley and of 1.3 m_0 in a satellite valley, which lies 0.37 eV above the Γ-point valley at $k = 0$. At low fields most of the electrons are in the Γ-point valley and have a mobility of $\sim$8,000 $\mathrm{cm^2/V\,s}$ (at room temperature). From here, electron heating is easily achieved at moderate fields and electrons are pumped into the satellite valley with a much lower mobility of $\sim$100 $\mathrm{cm^2/V\,s}$ causing a bias range of negative differential conductivity (Ridley and Watkins 1961; Hilsum 1962; see Fig. 4.1).

The field of directions for such mobility-induced negative differential conductivity shows characteristic differences to the previously discussed characteristics: while the quasi-neutrality curve now is field-independent, the drift current curve shows a minimum and an increase where the mobility decreases more than linearly with the field, and finally a horizontal range where the

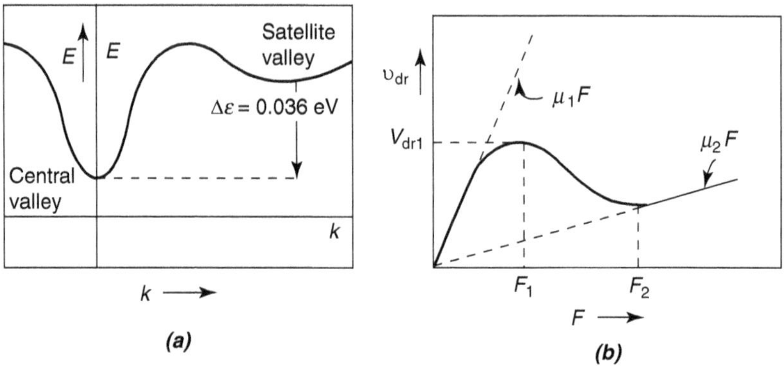

Fig. 4.1. (**a**) $E(k)$ of GaAs exhibiting a low-effective mass central valley and a high-effective mass satellite valley at 0.035 eV higher energy. (**b**) Drift velocity as function of the field, showing a negative differential conductivity range for $F_1 < F < F_2$

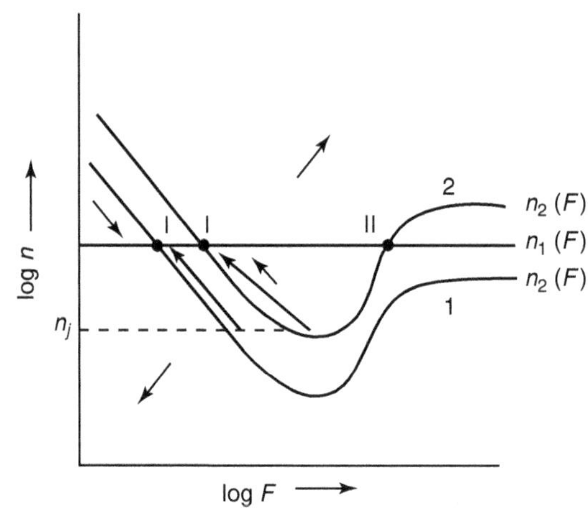

Fig. 4.2. Field of directions for constant carrier density and field-dependent mobility with quasi-neutrality curve $n_1(F)$ and drift current curves for low (curve 1) and high current (curve 2)

drift current saturates. When starting at the boundary concentration n_c, one loses the possibility of drawing a stationary solution when the minimum of the $n_2(F)$ curve crosses n_c. This can be seen from Fig. 4.2, as the solution curve would then have to slope away from the singular point I, thereby *prohibiting* a "long" solution curve, i.e., approach close to the singular point I to fill the entire device. One consequence is that, with further increased bias, the current can no longer increase[1] (i.e., current saturation is reached); however,

[1] The starting point of the solution would otherwise cross into the third quadrant with no possibility of reaching the singular point I necessary to obtain a solution with a bulk part ($n = \text{const}$, $F = \text{const}$).

the additional voltage drop cannot be absorbed since the Schottky barrier cannot convert into a domain[2] which in turn could expand. Consequently, the solution has to become nonstationary, as will be explained in Sect. 4.1.1.

4.1.1 The Gunn Effect

In 1963, Gunn reported microwave current oscillations through a doped GaAs sample under d.c. voltage bias (Gunn 1963). Furthermore, the applied voltage for which these oscillations appeared was found to coincide with the onset of a negative differential current regime within the current–voltage curves. The oscillations were later established to be due to moving domains of high-electric field that transits the sample with a speed very close to the saturation drift velocity for electrons in GaAs. The measurement was done in a set of classic scanning capacitive probe technique (Gunn 1965). In the case of the Gunn effect, the origin of negative differential conductivity is related to the k-space transfer of hot electrons from the high-mobility Γ point to the relatively low-mobility L point in the Brillouin zone. The domains can then be treated using the field-of-direction analysis similar to the one discussed earlier, but appropriately modified.

Field-of-Direction Analysis of Gunn-Domains

We will now use the field-of-direction analysis in finding domain-type solutions by stepwise relaxing of the conditions which prevented stationary domains of the type discussed earlier Fig. 4.3.

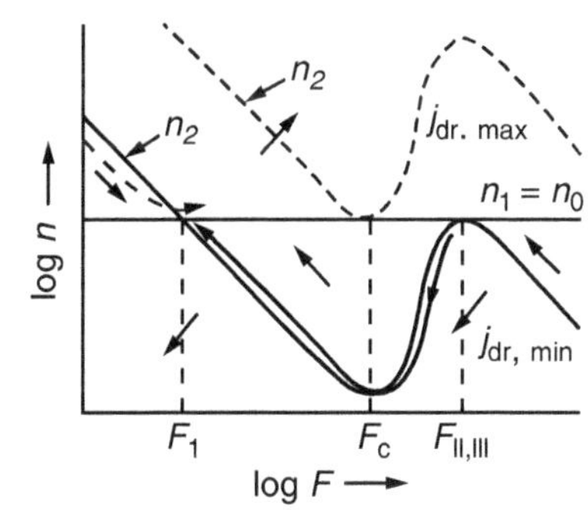

Fig. 4.3. Field of direction diagram with charge neutrality curve n_1 and diffusion neutrality curve n_2 for minimum $j_{n,min}$ and maximum current $j_{n,max}$ that support domain-type solution, one of which is identified by heavy arrow starting from a point near n_0, F_{II}

[2] The second singular point can never be reached by a stationary solution as long as n has to start at n_c.

When relaxing the requirement of connecting to the cathode, i.e., when starting at an electron density close to the bulk (or using a neutral or injecting cathode) one loses the ability of obtaining stationary high-field domains.

Moving Domain-Type Solutions

To maintain the ability to use the field of directions analysis, we will again analyze only undeformed domains that move with constant velocity, as discussed in Sect. 3.2.1. The corresponding set of transport and Poisson equations, here with a field-dependent mobility, is given by

$$\frac{\mathrm{d}n}{\mathrm{d}x'} = \frac{e}{\mu_\mathrm{n} kT}\left[n\mu_\mathrm{n}F - \frac{j}{e} - c(n - n_\mathrm{o})\right] \tag{4.1}$$

$$\frac{\mathrm{d}F}{\mathrm{d}x'} = \frac{e}{\varepsilon\varepsilon_0}(n - n_\mathrm{o}) \tag{4.2}$$

with n_o the constant electron density in the bulk (for charge neutrality) and c the domain velocity.

The charge neutrality curve n_1 in this model is independent of the field, while the diffusion neutrality curve $n_2(F)$ is non-monotonic and depends via $\mu_n(F)$ on the field (Fig. 4.4) and also is a sensitive function of the domain velocity c:

$$n_2(F) = \frac{j - ecN_0}{e[\mu_n(F)F - c]}. \tag{4.3}$$

Since $n_2(F)$ must remain positive, (4.3) has physically meaningful solutions only in the range in which $j - ecN_0$ has the same sign as $\mu_\mathrm{n}F - c$. These solutions are shown in Fig. 4.5 for different relations of the domain velocity to the drift velocity $\mu_\mathrm{n}F = v_0$.

Linearization around the three singular points again reveals that points I and III are saddle points and only circling around point II is possible for

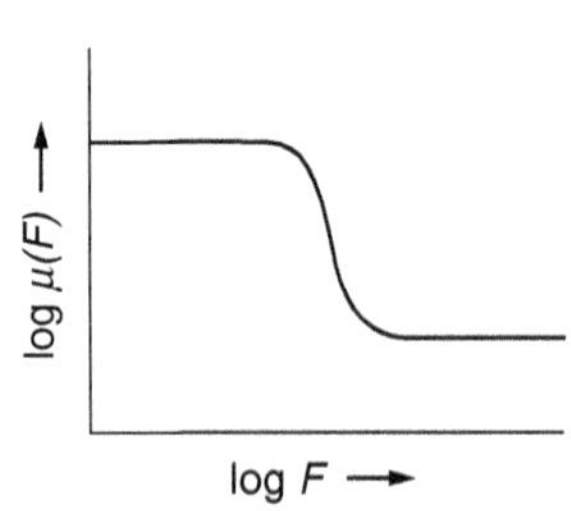

Fig. 4.4. Schematics of the field-dependent mobility assumed for the discussion in this section

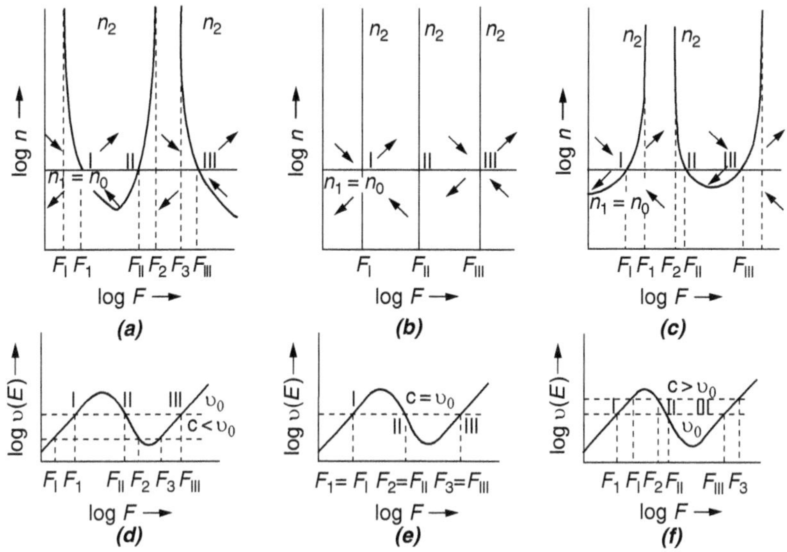

Fig. 4.5. Field of direction diagram with n_1 and $n_2(F)$ obtained from (4.1) and (4.2) for three different domain velocities, as identified in the lower part of the figure where the drift-velocity dependence on the field is shown.(**a**),(**d**), for $c < v_0$; (**b**),(**e**) for $c = v_0$; and (**c**),(**f**) for $c > v_0$

periodic solutions. The only permitted[3] periodic solution is that for a vortex with

$$c = \mu_n F_{\mathrm{II}}, \tag{4.5}$$

and with F_{II} determined by the total current. This means that an infinite number of periodic solutions exists that fulfill (4.5), depending on j; however, only one of these solution curves approaches the singular point I closely enough to permit a low and essentially constant bulk field and one moving domain (Fig. 4.6).

This result, as directly and convincingly obtained from the field of direction method demonstrates the usefulness of this method, and with proper parameters permits the determination of the bulk and domain field F_{I} and F_{T}, respectively.

---

[3] A permitted solution must fulfill (Butcher et al. 1966)

$$\int_{F_{\mathrm{L}}}^{F_{\mathrm{H}}} \left[v(F') - c\right] \mathrm{d}F' = 0, \tag{4.4}$$

i.e., it must cut of the $v(F)$ curve equal areas above and below v_0 as shown in Fig. 4.6b.

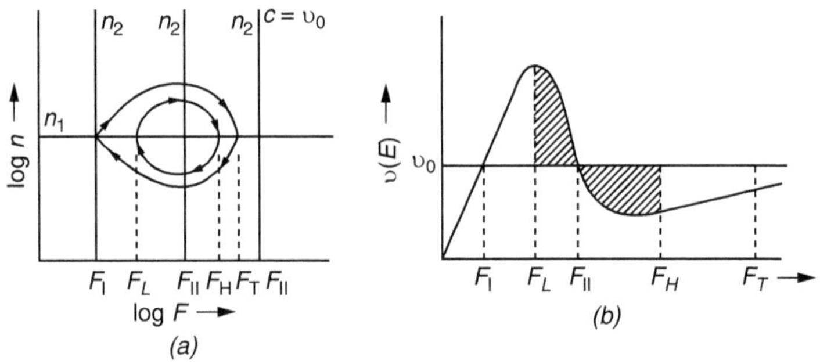

Fig. 4.6. (**a**) Field of direction as in Fig. 4.5b with two solution curves, one of them extending close to the singular point I for a long solution in the bulk. (**b**) Drift velocity distribution with low (F_L) and high field (F_H) identified, indicating equal areas above and below v_0. Field extrema must lie on n_1 (see (4.2))

Current Oscillations

When a high-field domain travels through the bulk of the device, the current tends to attain the lowest possible value compatible with the moving domain (minimum entropy condition). As soon as the domain reaches the anode, it disappears and the current increases. Thereby $n_1(F)$ and with it the singular point I are shifted toward higher fields (Fig. 4.3), bringing I closer to the point II, thereby facilitating the initiation of the next domain close to the cathode. After initiating is completed, the domain relaxes to a form that can move as an undeformed domain, thereby reducing the current again, and the process repeats itself, periodically.

Again, this conclusion of the dynamics of the domain creation and motion can best be seen by analyzing the field-of-direction, as done in the previous sections.

4.2 Summary and Emphasis

When the negative differential conductivity is caused by a stronger than linear decrease of the mobility with the electric field, high-field domain must occur that move trough the crystal and cannot remain attached to an electrode as a stationary domain. Such decrease of electron mobility with the field occur in GaAs as a result of conduction electron excitation into a higher conduction band with substantially lower effective mass. The resulting high-field domains in GaAs are not slowed down by carrier trapping and therefore move at high velocities given by the drift velocity of electrons. The consequent current oscillations are in the gigahertz range and can be used for high-frequency current generation (Gunn effect).

Gunn effect current oscillations caused by a decrease of the mobility with the electric field are of high technical interest as high frequency current generators.

5

Negative Differential Conductivity in Other Materials

Summary. It is instructive to discuss the development of high-field domains in a more general fashion and apply the results to a variety of other examples.

We will now discuss the solution curves of the transport equations in a more general approach. The discussion in this chapter is entirely based on a contribution of Schöll (Schöll 2002, Chap. 33). In bulk semiconductors, the drift current density of electrons can generally be written as

$$j(F) = en(F)v(F). \tag{5.1}$$

Here, e is the electron charge, $n(F)$ and $v(F)$ are the field-dependent electron concentration and drift velocity, respectively. The differential conductivity, assuming a homogeneous electric field, is

$$\sigma_{\text{diff}} = \frac{\mathrm{d}j}{\mathrm{d}F} = e\left(n\frac{\mathrm{d}v}{\mathrm{d}F} + v\frac{\mathrm{d}n}{\mathrm{d}F}\right). \tag{5.2}$$

There are then two distinct ways to achieve $\sigma_{\text{diff}} < 0$, i.e., negative differential conductivity:

(i) $\mathrm{d}v/\mathrm{d}F < 0$ which corresponds to negative differential mobility (*drift instability*) and
(ii) $\mathrm{d}n/\mathrm{d}F < 0$ which corresponds to negative differential carrier concentration (*generation–recombination instability*)

We explore some typical example systems below.

Two important classes of negative differential conductivity (NDC) are described by an N-shaped or an S-shaped $j(F)$ characteristic, as discussed earlier and now denoted by $NNDC$ and $SNDC$, respectively. However, more complicated forms like Z-shaped, loop-shaped, or disconnected characteristics are also possible, and may be treated analogously (Wacker and Schöll 1995). $NNDC$ and $SNDC$ are associated with voltage- or current-controlled instabilities, respectively. In the $NNDC$ case, the current density is a single-valued function of the field, but the field is multivalued: the $F(j)$ relation has three branches in a certain range of j. The *SNDC* case is complementary in the sense that F and j are interchanged. This duality is in fact far-reaching (Shaw et al. 1992).

The *global* current–voltage characteristic $I(V)$ of a semiconductor can, in principle, be calculated from the *local* $j(F)$ relation by integrating the current density j over the cross-section A of the current flow

$$I = \int_A \mathbf{j} \mathrm{d}\mathbf{f} \tag{5.3}$$

and the electric field F over the length L of the sample

$$V = \int_0^L F \mathrm{d}x. \tag{5.4}$$

Unlike the $j(F)$ relation, the $I(V)$ characteristic is not only a property of the semiconductor material, but also depends on the geometry, the boundary conditions, and the contacts of the sample. Only for the idealized case of spatially homogeneous states, the $j(F)$ and the $I(V)$ characteristics are identical, up to re-scaling. The $I(v)$ relation is said to display *negative differential conductance* if

$$\frac{\mathrm{d}I}{\mathrm{d}V} < 0. \tag{5.5}$$

In case of *NNDC*, the *NDC* branch is often but not always – depending upon external circuit, contacts, and boundary conditions – unstable against the formation of electric field domains, while in the *SNDC* case current filamentation generally occurs (Ridley 1963). These primary self-organized spatial patterns may themselves become unstable in secondary bifurcation leading to periodically or chaotically breathing, rocking, moving, or spiking filaments or domains, or even solid-state turbulence and spatiotemporal chaos (Schöll et al. 1998; Schöll 2001).

At this point **a word of warning** is indicated.

First, negative differential conductivity does not always imply instability of the steady state, and positive differential conductivity does not always imply stability. For example, *SNDC* states can be stabilized (and experimentally observed!) by a heavily loaded circuit (with a sufficiently high series resistance), and, on the other hand, the bifurcation of a self-sustained oscillation can occur on a $j(F)$ characteristic with positive differential conductivity.

Second, there is no one-to-one correspondence between *SNDC* and filaments, or between *NNDC* and domains.

Finally, it is important to distinguish between the local $j(F)$ characteristic and the global $I(V)$ relation which may exhibit negative differential conductance even if the $j(F)$ relation does not show *NDC*, and vice versa (Schöll 2000).

Domains and filaments can be theoretically described as special nonuniform solutions of the basic semiconductor transport equations, subject to appropriate boundary conditions. A linear stability analysis around the spatially homogeneous steady state (fixed or singular point) for small space- and time-dependent fluctuations of the electromagnetic field and the relevant transport variables, e.g., carrier concentrations or charge densities, yields conditions for the onset of domain-type or filamentary instabilities (Schöll 1987).

In the following sections of this chapter, we will analyze high-field domains.

5.1 Domain Instability

The formation of field domains in a sample with $NNDC$ can be understood from a simple argument. By way of example, we consider the case where the negative differential conductivity results from negative differential mobility, as in the Gunn effect (Gunn 1963; Gunn 1964), so that a range of fields exists where $\mathrm{d}v/\mathrm{d}F < 0$.

To illustrate the response of an $NNDC$ element to a charge fluctuation, consider a uniform field with a domain of increased field in the center of the sample as shown in Fig. 5.1a. The charge distribution that produces this field is shown in Fig. 5.1b. There is a net accumulation of charge on the left side of the domain and a depletion layer on the right. If we consider *negatively*

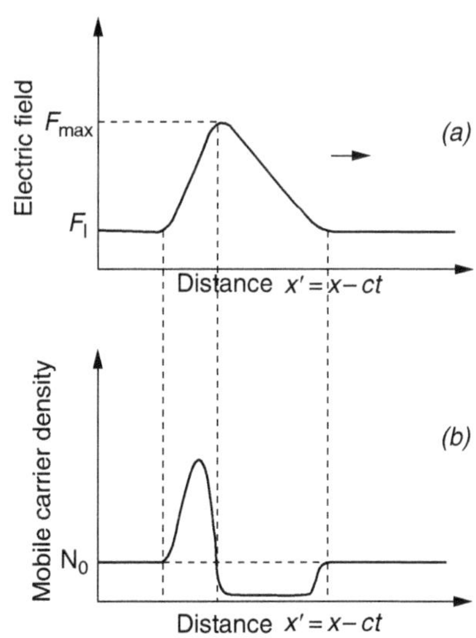

Fig. 5.1. (**a**) Electric field profile, (**b**) carrier density profile of a high-field Gunn domain. The domain is moving with velocity v in the positive x-direction (Schematic) (After Shaw et al. 1992)

charged carriers, the carriers and hence the domain will be moving to the right (cathode at left, anode at right).

Assuming that the field within the domain is within the NDC range and the field outside the domain is within the Ohmic range, but close to the field of peak velocity, then it is clear that the field fluctuation will initially grow with time. This happens because the higher upstream field in the center of the domain results in carriers moving more slowly than those at the edges, where the field is lower. Charge will therefore deplete on the right (leading) edge of the domain and accumulate at the left (trailing) edge. This charge will add to what is already there, increasing the field in the domain. If the element is in a resistive circuit the increasing voltage across the domain will cause the current to decrease in the circuit and lower the field outside the domain. The field in the domain will continue to grow in the interior of the domain and drop outside. Thus, an instability of the uniform field with respect to domain formation results.

The constitutive equations in one spatial dimension are the carrier continuity equation, neglecting carrier trapping and recombination processes,

$$\frac{\partial n(x,t)}{\partial t} + \frac{\partial}{\partial x}\left(n(x,t)v(F) - D\frac{\partial n}{\partial x}\right) = 0 \tag{5.6}$$

and Gauss' Law (Poisson equation)

$$\frac{\partial F(x,t)}{\partial x} = \frac{e}{\varepsilon\varepsilon_0}\left(n(x,t) - N_0\right), \tag{5.7}$$

where eN_0 is the negative uniform background charge density, and D is the diffusion constant. Equations (5.6) and (5.7) can be combined by differentiating (5.7) with respect to time and substituting into (5.7). Integration over x gives the total current density $j(t)$ which is composed of displacement current, drift, and diffusion contributions, respectively:

$$\varepsilon\varepsilon_0\frac{\partial F(x,t)}{\partial t} + e\left(n(x,t)v(F) - D\frac{\partial n}{\partial x}\right) = j(t) \tag{5.8}$$

with external current density $j(t)$. Substituting (5.7) into (5.8) to eliminate n yields the governing nonlinear transport equation

$$\varepsilon\varepsilon_0\left(\frac{\partial F(x,t)}{\partial t} + v(F)\frac{\partial F}{\partial x} - D\frac{\partial^2 F}{\partial x^2}\right) + eN_0 v(F) = j(t). \tag{5.9}$$

The stability of the uniform steady-state F^* can be tested by linearizing (5.9) around F^* for small space and time dependent fluctuations

$$\delta F(x,t) = \exp(\lambda t)\exp(ikx). \tag{5.10}$$

Substituting this ansatz into (5.9) for fixed external current density j determines the time increment λ for any given wave vector k, i.e., the dispersion relation

$$\lambda = -\frac{1}{\tau_M} - Dk^2 - \mathrm{i}kv(F^*). \tag{5.11}$$

Here, the effective differential dielectric relaxation time

$$\tau_M \equiv \left(e\varepsilon\varepsilon_0 N_0 \frac{\mathrm{d}v}{\mathrm{d}F} \right)^{-1} \tag{5.12}$$

has been introduced. For negative differential mobility ($\mathrm{d}v/\mathrm{d}F < 0$), τ_M is negative, and small fluctuations

$$\delta F(x,t) = \exp\left[-t \left(\frac{1}{\tau_M} + Dk^2 \right) \right] \exp\left(ik(x - v_0 t) \right) \tag{5.13}$$

grow in time for $k^2 < (D|\tau_M|)^{-1}$, i.e., for long wavelengths $2\pi/k$. Equation (5.13) describes an undamped traveling wave propagating in the positive x direction with velocity $v_0 = v(F_0)$.

This instability leads to the bifurcation of moving field domains from the uniform steady state, which depends, however, sensitively upon the sample length, the boundary conditions at the cathode contact and the external circuit (Shaw et al. 1979, 1992). A simple criterion for the onset of a moving field domain can be derived from (5.13) by neglecting diffusion and assuming that the linearization remains valid throughout its transit across the element of length L. The growth at the end of the transit, i.e., after the transit time $t = L/v_0$, is then given by

$$G = \frac{\delta F(t)}{\delta F(0)} = \exp\left(\frac{L}{v_0|\tau_M|} \right) = \exp\left(N_0 Le \left| \frac{\mathrm{d}v}{\mathrm{d}F} \right| \varepsilon\varepsilon_0 v_0 \right). \tag{5.14}$$

Substantial growth, i.e., a moving domain instability, occurs if the exponent in (5.14) is larger than unity, or

$$N_0 L > \frac{\varepsilon\varepsilon_0 v_0}{e\left|\frac{\mathrm{d}v}{\mathrm{d}F}\right|}. \tag{5.15}$$

This is the $N_0 L$ product stability criterion for the Gunn effect derived by McCumber and Chynoweth (1966), Kroemer (1968).

The fully developed field domains must be calculated from the full nonlinear transport equation (5.9), subject to appropriate boundary conditions. A simple visualization of the moving domain solution can be obtained by transforming (5.9) to the co-moving frame $x' = x - ct$, where c is the domain velocity, yielding:

$$\varepsilon\varepsilon_0 \left([v(F) - c] \frac{\partial F}{\partial x'} - D \frac{\partial^2 F}{\partial x'^2} \right) + eN_0[v(F) - v_\infty] = 0. \tag{5.16}$$

Here, $j = eN_0 v_\infty$ with $v_\infty = v(F_\infty)$ is the current density in the neutral material outside the domain. The result can be written in the form of a "dynamic system" of two first-order differential equations for F and $n = (\varepsilon\varepsilon_0/e)\partial F/\partial x + N_0$, where the parameter corresponding to "time" is given by x'.

The topology of the solutions may be conveniently discussed in terms of a *phase portrait analysis* or *field-of-directions analysis* (Shaw et al. 1992) which gives qualitative insight into the nature of the solutions and the effect of boundary conditions.

Such methods have been extensively used by Böer et al. (1967, 1968, 1969) in the early investigations of Gunn domains and recombination domains, and reviewed in Sect. 4.2.

Further, it turns out that the triangular shape of the Gunn domain depicted in Fig. 5.1 can be understood on the basis of an equal areas rule (Butcher 1965, Knight and Peterson 1967) which can be derived from (5.16), see Fig. 6.18 and discussion below. It determines, for a given domain velocity $c = v(F_\infty)$, the peak field $F_{\max}$ of the domain. It should be noted that the low-field state F_∞ corresponds to a homogeneous steady state (fixed point), but the peak field $F_{\max}$ does not. Therefore, the second rising branch of the $NNDC$ characteristic is not necessary for the existence of the domain: The Gunn domain does not involve phase coexistence of two bistable homogeneous states.

Note that more elaborate mathematical analyzes, including realistic boundary conditions, establishing the existence and stability of traveling domain solutions, have also been given (Higuera and Bonilla 1992; Bonilla et al. 1994, 1997).

Equal areas rules have been derived as simple geometrical relations which provide some relevant information about the domains or filaments, like the peak field and the propagation velocity of the domains, or the filament radius as a function of the material parameters and the applied bias, for a variety of cases (Schöll and Landsberg 1988). Extensions to field domains in superlattices (Schwarz and Schöll 1996; Wacker et al. 1997) have also been given.

5.2 Field Domains in Superlattices

Vertical high-field transport in GaAs/AlAs super lattices has been shown experimentally (Esaki and Chang 1974; Kawamura et al. 1986; Choi et al. 1987; Helm et al. 1989; Helgesen and Finstad 1990; Grahn et al. 1991; Zhang et al. 1994; Merlin et al. 1995; Kwok et al. 1995; Mityagin et al. 1997) to be associated with negative differential conductivities and field domain formation induced by resonant tunneling between adjacent quantum wells.

The field domains may either be stationary, leading to characteristic sawtooth current–voltage characteristics (Esaki and Chang 1994) or traveling, associated with self-sustained current oscillations (Kastrup et al. 1995; Hofbeck et al. 1996). In strongly coupled superlattices, i.e., superlattices with small barrier widths, oscillations above 100 GHz at room temperature

(Schomburg et al. 1998, 1999) have been observed experimentally, while in weakly coupled superlattices the frequencies are many orders of magnitude lower (Kastrup et al. 1997). In the former case, the impact of electron heating may cause S-shaped negative differential currents in addition to the N-shaped negative differential currents occurring at lower fields, and the combination of S- and N-type instabilities then leads to a modified structure of the high-field domains associated with self-generated gigahertz oscillations (Steuer et al. 1999, 2000). The existence of domain states and of multistable, sawtooth-like current–voltage characteristics consisting of many branches has been explained for structurally perfect, weakly coupled samples by phenomenological (Korotkov et al. 1993; Miller and Laikhtman 1994; Prengel et al. 1994; Bonilla et al. 1994) or microscopic (Wacker 1998; Cao and Lei 1999) models. The standard approaches of miniband transport, Wannier–Stark hopping, and sequential tunneling have been derived from a full quantum transport theory using the framework of nonequilibrium Green functions as limit cases depending on the mutual ratios of three energy scales (Wacker and Jauho 1998; Wacker 2002):

1. The miniband width as a measure for the coupling between the wells,
2. The potential drop across one superlattice period and
3. The scattering width.

It has also been found in computer simulations that small amounts of imperfections and disorder, such as fluctuations of doping, well or barrier widths, sensitively influence both the stationary domains (Wacker and Schöll 1995; Schwarz and Schöll 1996; Schwarz et al. 1996) and the domain oscillations (Patra et al. 1998; Schwarz et al. 1998; Schöll et al. 1998; Steuer et al. 1999).

Experimentally, the global properties have been directly related to the extent of disorder by X-ray analysis of superlattices (Grenzer et al. 1998). Chaotic dynamics has also been found both theoretically (Bulashenko et al. 1995, 1996; Bonilla et al. 1996) and experimentally (Zhang et al. 1996; Luo et al. 1998; Bulaschenko et al. 1999), if an ac driving bias is applied.

For dc driven superlattices chaotic behavior has also been demonstrated (Amann et al. 2002a). For appropriate boundary conditions at the emitter contact, self-sustained dipole waves are spontaneously generated at the emitter. The dipole waves are associated with traveling field domains, and consist of electron accumulation and depletion fronts which in general travel at different velocities and may merge and annihilate. Depending on the applied voltage and the contact conductivity, this gives rise to various oscillation modes as well as different routes to chaotic behavior.

Let us now briefly discuss domain formation in doped or undoped, photo excited GaAs/AlAs superlattices (e.g., Grahn et al. 1995; Wacker 2002). The data show static domain walls, oscillating domain walls, as well as propagating domain walls. Here, the term domain wall refers to the region of the superlattice that separates the low-field side of the sample from the high-field side of the sample.

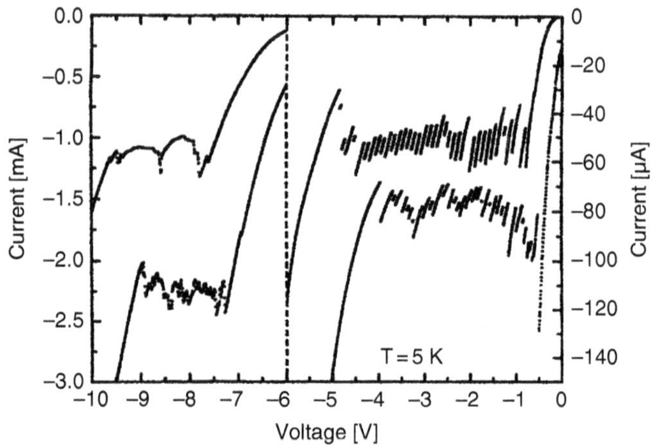

Fig. 5.2. Measured current–voltage characteristics of a GaAs/AlAs superlattice at 5 K. The upper curve was recorded without illumination, the lower one with illumination. Note the different vertical scale in the left- and right-hand parts, respectively (After Grahn et al. 1995)

An experimental example of static domain walls is shown in Fig. 5.2 on the right-hand side. Each of these branches in the current–voltage curve corresponds to the domain wall lying in one particular period (quantum well) of the superlattice. When the current jumps to the adjacent step, then the corresponding domain wall jumps to the next period in the superlattice. Experimental results for this system are well-described by phenomenological theories that appear in the form of a discrete drift–diffusion model. The key nonlinearity arises from the field dependence of electron tunneling from one period of the superlattice to the adjacent one (Schöll 2001). The irregular structure of the different branches in the current–voltage characteristic can be modeled in good agreement with the experiment by including doping or well width or barrier width fluctuations in the simulations (Wacker and Schöll 1995; Schwarz et al. 1996). If the carrier density in the wells is not high enough to provide the space charge necessary to form a stable, stationary domain wall, then oscillations may arise. On the left-hand side of Fig. 5.2 temporal current oscillations occur as shown in the time trace of Fig. 5.3a. Figure 5.3b depicts the spectrally and temporally resolved photoluminescence intensity which maps the space–time dynamics of the field domains, since the high-field domain corresponds to a lower luminescence energy than the low-field state due to the quantum confined Stark effect. It can be clearly seen that the high-field domain appears periodically.

Figure 5.4 shows simulations of current oscillations in a strongly coupled superlattice, where the frequencies are much higher. The space–time plot and the current trace depict traveling field domains and non-sinusoidal oscillations.

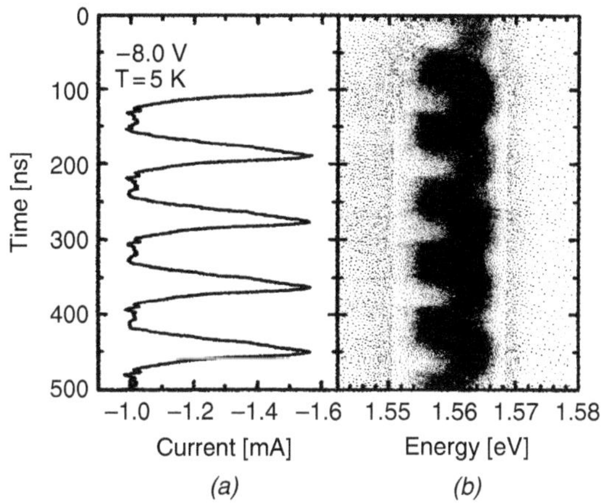

Fig. 5.3. (**a**) Typical measured time traces of the time-resolved current for an applied voltage of −8 V for the superlattice of Fig. 5.2. (**b**) Time-resolved photoluminescence intensity vs. photon energy and time showing domain oscillations. Darker regions correspond to a larger intensity (After Grahn et al. 1995)

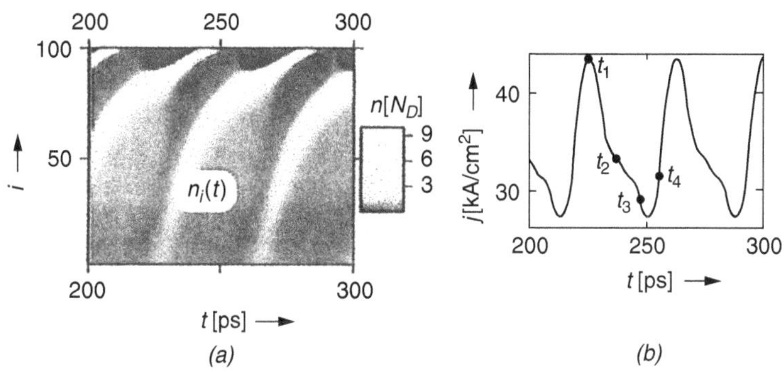

Fig. 5.4. Calculated oscillations of high-field domains in a strongly coupled superlattice. (**a**) Electron density vs. position i and time t. Brighter regions correspond to higher electron densities. (**b**) current density vs. time. (After Steuer et al. 2000)

The dipole domain propagation in a superlattice can be controlled by a high-frequency driving field (Schomburg et al. 2002). The high-frequency impedance of a driven superlattice is also strongly influenced by moving domains (Jappsen et al. 2002). Chaotic behavior can be simply suppressed by a time-delayed feedback loop in the control circuit (Schlesner et al. 2003), which stabilizes unstable periodic orbits embedded in the chaotic attractor (Schöll et al. 2008).

Theoretical and experimental research has recently shown that noise can have surprisingly constructive effects in many nonlinear systems. In particular, an optimal noise level may give rise to ordered behavior and even produce new dynamical states. In spite of considerable progress on a fundamental level, useful applications of noise-induced phenomena in technologically relevant devices are still scarce. Recently, it has been demonstrated that noise can give rise to moving field domains in semiconductor superlattices even if the deterministic system allows only for a stationary domain state, and that these space–time patterns can be controlled by the time-delayed feedback scheme applied to purely deterministic chaotic front patterns in a superlattice previously.

It has also been clarified how the transition from stationary domains to moving field domains occurs in a superlattice (Hizanidis et al. 2006). This scenario corresponds to a global bifurcation on a limit cycle which is called *saddle-node infinite period bifurcation or SNIPER*, since the period of the oscillations tends to infinity as the bifurcation point is approached. Keeping, e.g., the contact conductivity fixed and increasing the voltage, a limit cycle of approximately constant amplitude and increasing frequency is born. This happens through the collision of a stable fixed point (node) and a saddle-point.

If the system is prepared in the stable stationary domain state, which corresponds to a stationary charge accumulation front, and noise is introduced, then with increasing noise intensity the behavior of the system changes dramatically: the accumulation front remains stationary only for a while, until a pair of a depletion and another accumulation front (i.e., a charge dipole with a high-field domain in between) is generated at the emitter. As is known from the deterministic system, this dipole injection depends critically upon the emitter current. Here, it is triggered by noise at the emitter. Because of the global voltage constraint of a fixed voltage applied to the superlattice, the growing dipole field domain between the injected depletion and accumulation fronts requires the high-field domain between the stationary accumulation front and the collector to shrink, and hence that accumulation front starts moving toward the collector. For a short time, there are two accumulation fronts and one depletion front in the sample, thereby forming a tripole (Amann et al. 2002), until the first accumulation front reaches the collector and disappears. When the depletion front reaches the collector, the remaining accumulation front must stop moving because of the global constraint, and this happens at the position where the first accumulation front was initially localized. After some time, noise generates another dipole at the emitter and the same scenario is repeated.

There are two distinct time scales in the system. One is related to the time the depletion front takes to travel through the superlattice. The other timescale is associated with the time needed for a new depletion front to be generated at the emitter. These two time scales are also visible in the noise-induced current oscillations. The time series of the current density are in the form of a pulse train with two characteristic times: the activation time, which is the time needed to excite the system from this stable fixed point (time needed for a new depletion front to be generated at the emitter) and the excursion time which is the time needed to return from the excited state to the fixed point (time the depletion front needs to travel through the device). Low noise is associated with large activation times and small, almost constant, excursion times, while as the noise level increases activation times become smaller and at sufficiently large noise intensity vanish. At low noise, the spike train looks irregular, and the interval between excitations is relatively large and random in time. At moderate noise, the spiking is rather regular therefore suggesting that the mean interspike-interval does not vary substantially. Further increase of noise results in a highly irregular spike train with very frequent spikes. This is the phenomenon of **coherence resonance**.

In the regime of stationary domains, switching between different branches of the multistable current–voltage characteristics and domain boundary relocation may be induced by voltage pulses as predicted theoretically (Amann et al. 2001) and verified experimentally (Rogozia et al. 2002). Such switching may not occur uniformly, and hence the coupled lateral and vertical electron dynamics should be considered. This amounts to complex combinations of moving field domains and current filaments (Amann et al. 2005; Xu et al. 2009).

5.3 Other Examples of Stationary and Moving High-Field Domains

Moving field domains due to the Gunn effect (Shaw et al. 1979, 1992) or a recombination instability (Bonch-Bruevich et al. 1975) have been extensively investigated experimentally and theoretically in the 1960s and 1970s. The earliest studies of such $NNDC$ and $SNDC$ instabilities in semiconductors were published by Böer et al. for CdS (1961, 1961) and Si (1961). They discovered moving high-field domains in CdS single crystals (Böer 1959), long before the observation of Gunn domain oscillations in GaAs were reported (Gunn 1963, 1964), and developed the electro-optical method described in Chap. 2 to measure these field inhomogeneities (Böer et al. 1959a) based on the first experimental evidence of the Franz–Keldysh effect.

Well-known mechanisms for field domain formation in bulk semiconductors are provided, e.g., by field-quenching in CdS (Böer et al. 1959) and in semi-insulating GaAs (Northrup et al. 1964; Samuilov 1995; Piazza et al. 1977), intervalley transfer (Ridley and Wallis 1961; Gunn 1963; Kroemer 1964), nonlinear generation–recombination kinetics in the regime of impurity impact-ionization breakdown (Kahn et al. 1991, 1992 a,b; Cantalapiedra et al. 1993; Bonilla et al. 1994, 1997, Bergman et al. 1996) in p-Ge, and the photo refractive Gunn effect (Segev et al. 1996; Bonilla et al. 1998).

More recently, field domains have also been studied in low-dimensional semiconductor structures. Bragg scattering in superlattices has been proposed as a mechanism for traveling domains in superlattices (Büttiker and Thomas 1977, 1978, 1979), and experimentally verified much later (Hofbeck et al. 1996). Periodically or chaotically moving domains induced by real-space transfer of electrons in modulation-doped semiconductor heterostructures have been predicted theoretically (Döttling and Schöll 1993, 1994; Borisov et al. 1999). While static negative differential mobility based on real-space transfer has been demonstrated theoretically (Gribnikov and Mel'nikov 1973; Hess et al. 1979) and found experimentally (Keever et al. 1981), the observed oscillatory behavior (Coleman et al. 1982; Balkan and Ridley 1989; Vickers et al. 1989) has not clearly been associated with moving field domains since spatially *uniform* oscillations have also been predicted from various models (Aoki et al. 1989; Schöll and Aoki 1991; Hendricks et al. 1991; Döttling and Schöll 1992).

Following the pioneering work of Esaki and Chang (1974) on resonant tunneling, field domains in weakly coupled superlattices have been extensively studied experimentally in the last decade (Kawamura et al. 1986; Choi et al. 1987; Helgesea and Finstad 1990; Grahn et al. 1991, 1991; Zhang et al. 1994; Merlin et al. 1995; Kwok et al. 1995; Mityagin et al. 1997). Both stationary and oscillating domains have been found and explained by various models (Korotov et al. 1993; Miller and Laikhtman 1994; Prengel et al. 1994; Bonilla et al. 1994; Wacker 1998, 2002 Cao and Lei 1999; Amann et al. 2005). If an ac driving bias is applied, moving domains exhibit particularly complex spatiotemporal dynamics, as simulations for Gunn diodes (Nakamura 1989; Mosekilde et al. 1990; Zongfu and Benkum 1991; Mosekilde et al. 1993), modulation-doped heterostructures Döttling and Schöll 1994), low-temperature impurity breakdown in p-Ge (Bergmann et al. 1996), and sequential resonant tunneling in superlattices (Bulashenko and Bonilla 1995; Bulashenko et al. 1996; Bonilla et al. 1996; Schomburg et al. 2002; Jappsen et al. 2002) have revealed.

As specific examples, we have discussed recombination instabilities in CdS and p-Ge (Sects. 3.2.1 and (5.3)) along with Ohmic boundary conditions at the injecting contact

$$F(x=0,t) = \rho_c j(t), \tag{5.17}$$

where ρ_c is the contact resistivity, and the global constraint for the applied dc voltage bias yield high-field domains similar to those observed in experiments (Kahn et al. 1991; Cantalapiedra et al. 1993). Figure 5.5 shows the simulation results, and Fig. 5.6 shows experimental results. Figure 5.7 shows the experimental current trace which should be compared with the simulated current trace in Fig. 5.5. Note that the current trace is non-sinusoidal which is to be expected when the underlying periodic structure is a localized traveling domain.

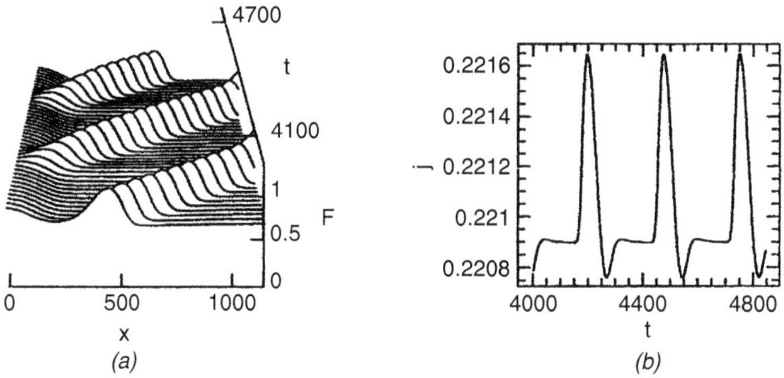

Fig. 5.5. Calculated high-field domains in *p*-Ge. In (**a**) the electric field is shown vs position and time; (**b**) corresponding current density $j(t)$ (all in dimensionless units) (After Cantalapiedra et al. 1993)

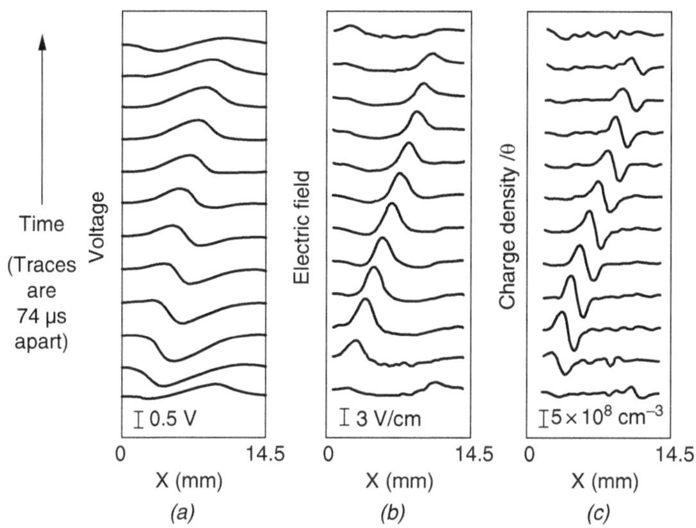

Fig. 5.6. Measured high-field domains in p-Ge. (**a**) Local sample voltage vs. position x at successive times with an applied bias of $F_{\mathrm{dc}} = 6.528\,\mathrm{V/cm}$. The injecting contact is at $x = 0$ and the receiving contact is at $x = 14.5\,\mathrm{mm}$. (**b**) Local electric field vs. position. (**c**) Local charge density vs. position (After Kahn et al. 1991)

5.4 Summary and Emphasis

When the mobility decreases steeper than linearly with the electric field high-field domains are observed that move from one to the other electrode. There are no stationary solutions possible for such high-field domains.

In GaAs, the reduction of electron mobility is caused by a redistribution of electrons over different bands within the conduction band, causing a substantial reduction of their effective mass. Such redistribution is a fast process

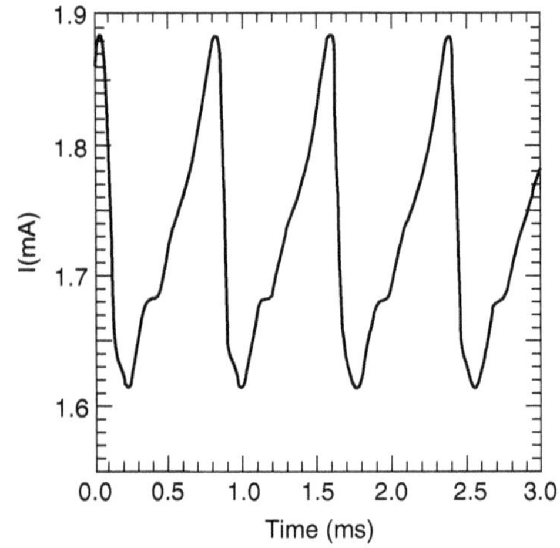

Fig. 5.7. Measured time traces of the total sample current corresponding to Fig. 5.6 (After Kahn et al. 1991)

and consequently the domains move at high frequency as long as they are not slowed down by space charge redistribution. The concurrent oscillation of the current is used in Gunn-diodes as a high-frequency generator.

The field-of-direction analysis can again be used advantageously to discuss the Gunn domains as they move with constant speed within the device, away from the electrodes.

There is a large variety of other cases in which negative differential conductivity regimes occur with consequent high-field domains or related instabilities that deserve a more general discussion.

Examples include high-field domains in superlattices, induced by resonant tunneling between adjacent quantum wells that have generated oscillations above 100 GHz. A number of experimental methods have been devised to investigate such domains including a more detailed analysis of the current–voltage characteristics and photoluminescence.

These instabilities can be discussed more generally in an analytical analysis of the transport and Poisson equations that is given in this chapter.

Moving high-field domains caused by a super-linear decrease in carrier mobility are of high technical interest for ac generation.

6

Current Channels

Summary. Current filaments are formed when the local characteristic of the current density vs. electric field or voltage is S-shaped such that at a given applied voltage there exist two locally stable states of low and high conductance, respectively.

In Chap. 5, we have discussed field-inhomogeneities that are induced by a variety of negative differential conductivity effects that result under dynamic observations in N-shaped current–voltage characteristics.

A different kind of inhomogeneity is observed in an *S-shaped negative differential conductance* range. Here, the *current density* becomes inhomogeneous; a current channel is formed. It consists of a channel of high conductance embedded in the bulk of low conductance. Thus, the current density is modulated in the plane perpendicular to the current flow as shown schematically in Fig. 6.1b. Typically, the high-conductance and the low-conductance states correspond to the upper and the lower branches of the S-shaped characteristic, respectively. The intermediate branch of negative differential conductance is generally unstable and the corresponding steady states break up into inhomogeneous current channel flow. Within the device, the temperature in its center will rise more than near its surface, and the center region will become more conductive. This produces even more Joule heating, causing a hot channel in the device with much higher current density in its center region (Böer et al. 1958, 1959, 1961; Böer and Döhler 1969 – for an example of filament modeling see Sect. 6.1.3). This current channel may be stationary or spatiotemporal[1] or, at high enough supplied power, if not stabilized by an external series resistor, may run-away into "thermal breakdown."

Since most of these effects develop in time, again we have resorted to a cinematographic documentation that is provided as a second part of the film on the Web site of the Springer Verlag. It is again highly instructive and also

[1] With substantial Joules heating, the current–voltage characteristic becomes S-shaped and requires stabilization with a sufficiently large series resistor to prevent a run-away thermal breakdown.

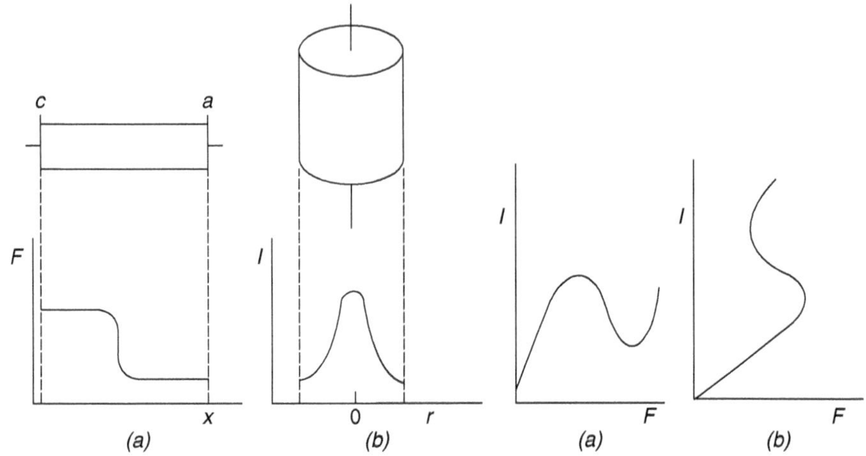

Fig. 6.1. (*left part*) Typical sample arrangement for investigating of field domains (**a**) and current channels (**b**) (*right part*) Schematics of N-shape (**a**) and stabilized S-shape current voltage characteristics (**b**)

contains an extensive educational part. Again, much of the detail that cannot be presented in a book that only provides single pictures is only visible in the film. We will refer from part to part in the book when this detail becomes important to the film.

6.1 Initiation of Current Channels

Current channels are usually initiated by a current node either in the center of a device with ohmic contacts, or close to a blocking electrode. In the latter case, a transition from a high-field domain to such a node may accompany an initiating process in some semiconductors that show such domains, e.g., photo conductive CdS. An example for such a rare domain type initiation that is fully developed is shown in Fig. 6.2. See also the film on the Web site.

During the motion of a high-field domain, the current steeply rises when the domain reaches (in an n-type semiconductor) the anode. If the current reaches values at its maximum that are sufficient for significant Joule heating, a new phenomenon sets in as a current filament is initiated, and oscillations will stop (Böer 1960). The initial phase of this transition from a moving high-field domain to a heated filament in the center part of the device can be seen in Fig. 6.3. Here, the domain (visible by Franz–Keldysh shift of the band edge) opens up and a darkening of the center part of the opening indicates a channel formation with a band-edge shift now due to a rise in temperature.

The further development of a current filament from a moving high-field domain is shown in Fig. 6.3. After the current channel is fully developed, the high-field domain is no longer observed.

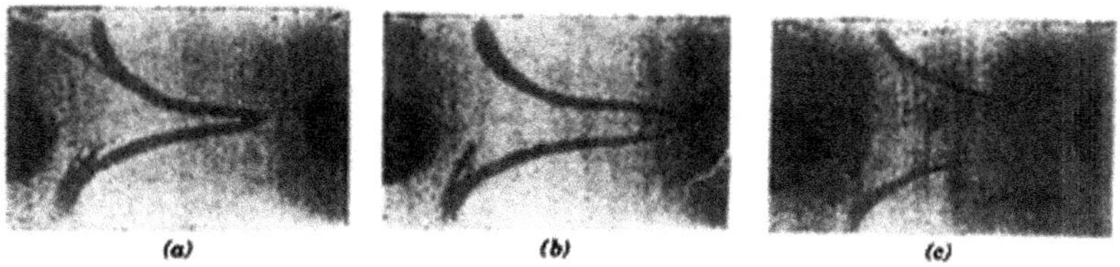

Fig. 6.2. Transition from a high-field domain after bulging in its center (**a**) because of some heating, to connecting to the anode (**b**) and opening up to a completely initiated current channel (**c**) with already visible heating (band-edge shift) in its center (After Böer 1960)

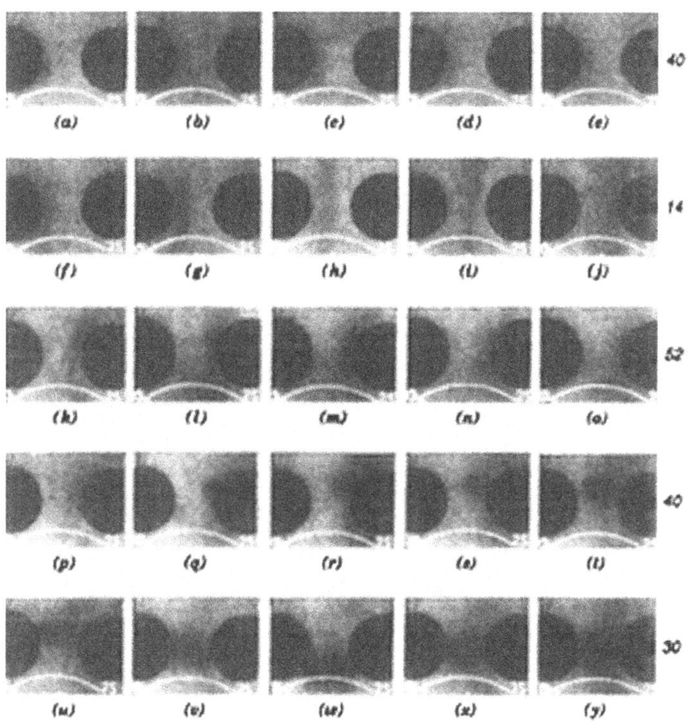

Fig. 6.3. Kinetics of the development of a moving high-field domain f–p followed by a node-formation at the anode (q), expanding toward the cathode until a current filament is formed (u) that moves up and down between the electrodes (v–x) until it becomes stationary (y). The number at the right border identifies the number of frames per row from which the five frames of that row were selected. The pictures were taken using the Franz Keldysh effect to make the field distribution visible, and the larger shift of the absorption edge due to Joule's heating. Cinephotography with 24 frames/s

6.1.1 Current Channels in Dielectrics and Devices

Current channels due to Joule heating were associated early on with a certain type of dielectric breakdown, the so-called thermal breakdown which occurs in dielectrics with sufficient residual conductivity that could lead at high enough electric fields to Joule heating (Wagner 1922). When embedded in other material that impedes heat dissipation, further heating could lead to a runaway current with consequent thermal material destruction, the *thermal breakdown*.

The electro-thermal balance between electrical energy input, Joule heating, and heat dissipation was later refined (see, e.g., Fock 1927; Whitehead 1939; Lueder and Spenke 1953; Skanavi 1958; Böer and Döhler 1969) and describes well the phenomenon of the thermal breakdown. For an early review, see Whitehead (1951) and Franz (1956).

Another example for electro-thermal channel formation are certain Ovonic switches (Fritzsche and Ovshinsky 1970; Böer and Ovshinsky 1970). These switches consist of a layer of an amorphous semiconducting glass between two graphite electrodes. Typical configuration and current–voltage characteristics of such a switch are given in Fig. 6.4. Near the threshold voltage, the current density in the central part of the device is high enough to initiate a current filament that in turn starts a transition along the loadline (2) to the holding branch (3). A return along another loadline (4) to the low-conducting branch (1) occurs when the current is decreased below the holding current. Part of the switching is electronic (Ovshinsky 1968). For the evaluation of the thermal part, see Böer et al. (1970).

6.1.2 Thermo-Optical Method to View Current Channels

The current channels can be made visible by a method similar to that used for the electro-optical detection by the Franz–Keldysh effect of the high-field

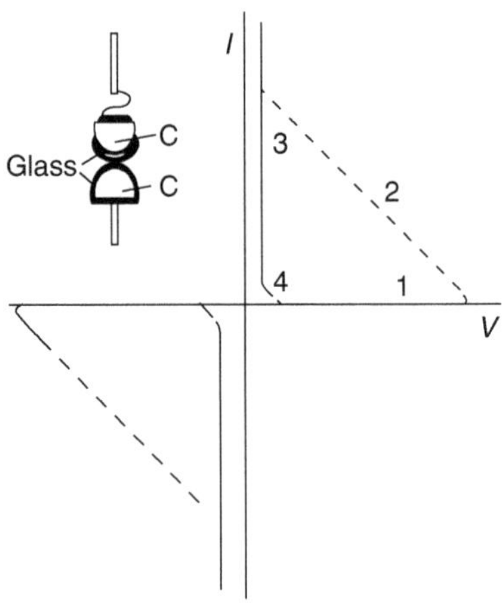

Fig. 6.4. (**a**) Ovonic threshold switch and (**b**) current–voltage characteristic with switching branches 2 and 4

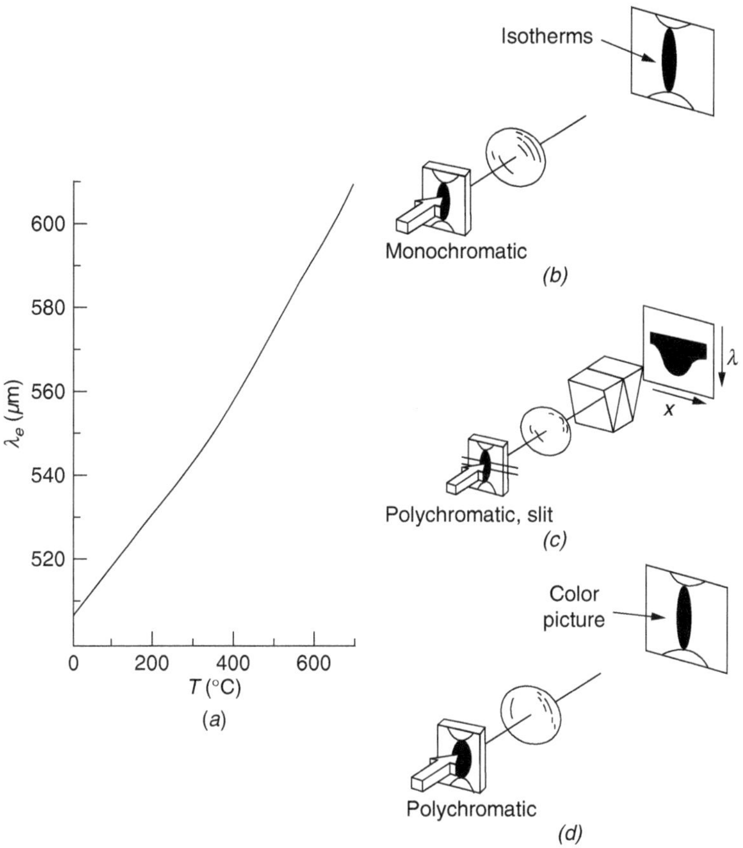

Fig. 6.5. Electro-thermo-optical method to visualize channel formation using the shift of the optical absorption edge with temperature. (**a**) Position of the band edge, shown at ($\alpha_o = 10^4\,\text{cm}^{-1}$) as function of the temperature in CdS. (**b**) Observation of an isotherm by using the appropriate monochromatic light. (**c**) Observation of the temperature distribution across a current filament. (**d**) Observation of a substantially heated current filament in polychromatic light (After Böer et al. 1961)

domains; however, here we use the shift of the absorption edge with increasing temperature. This shift is substantially larger for the temperatures attained in typical current filaments than the Franz–Keldysh effect in high-field domains. Therefore, monochromatic *or* polychromatic light can be used for such an analysis (Fig. 6.5).

Such current channels are shown in Fig. 6.6. They are usually initiated between the closest points of two electrodes. However, when moved[2] toward a narrow extension of one electrode (Fig. 6.6b), the filament is pushed out

[2] The filament can be moved by directional external cooling or in CdS by photochemical reactions (Böer and Borchardt 1954) at the appropriate temperatures,

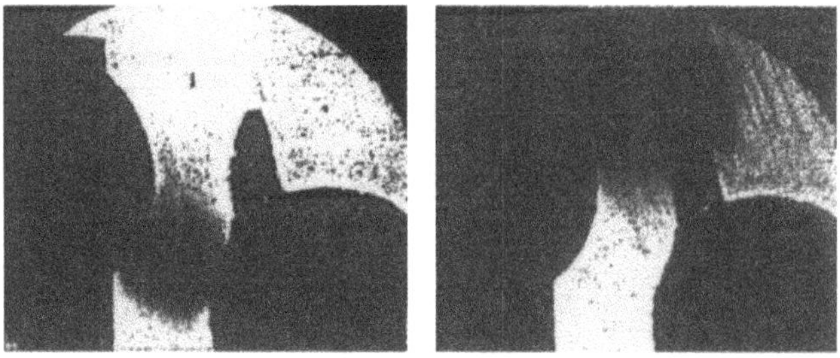

Fig. 6.6. Current filament made visible by the thermal shift of the optical absorption edge in CdS. (**a**) After initiation at sufficient bias the filament is established between the closest points between the left and lower right electrodes. (**b**) When the filament is moved to the neck of the extension of the right electrode, the filament moves rapidly to its tip by the magnetic field which tends to straighten the current path along the axis of the extended part of the electrode (After Böer 1960)

to the end of this extension and its current lines tend to follow the direction of the narrow electrode, indicating an influence of the magnetic field in forming the actual shape of the filament by its tendency to form a straight current path in continuation of the extended electrode axis. Again, for more detail see the film at the Web site of Springer.

6.1.3 Current Channel Modeling

A rather simple and transparent description can be given for the development of a current channel in the center axis of an idealized semiconducting cylinder (see Fig. 6.1b), which we will now present.

For the description of the major channel features (Böer 1971), we will neglect here space charge effects, thus describing the current by drift alone:

$$\mathbf{j}_n = \sigma_n \mathbf{F}. \tag{6.1}$$

The temperature-dependent conductivity $\sigma_\mathrm{n} = e\mu(T)n(T)$ is essentially determined by the exponential temperature dependence of n (neglecting the lesser temperature dependence of the mobility):

$$\sigma_\mathrm{n} \simeq e\mu_0 n_0 \exp\left(-\frac{E_\mathrm{c} - E_\mathrm{Fn}}{kT}\right) = \sigma_0 \exp\left(-\frac{E_\mathrm{c} - E_\mathrm{Fn}}{kT}\right). \tag{6.2}$$

The interaction between heat production by the current and heat dissipation to the surroundings is given by the thermal balance equation

$$c_\mathrm{p}\frac{\partial T}{\partial t} = \sigma(\mathbf{grad}\,\psi)^2 + div(\kappa\,\mathbf{grad}\,T). \tag{6.3}$$

which cause a reduction of the photo current in the heated regions and initiate a movement of the filament into regions not yet heated (Böer 1960).

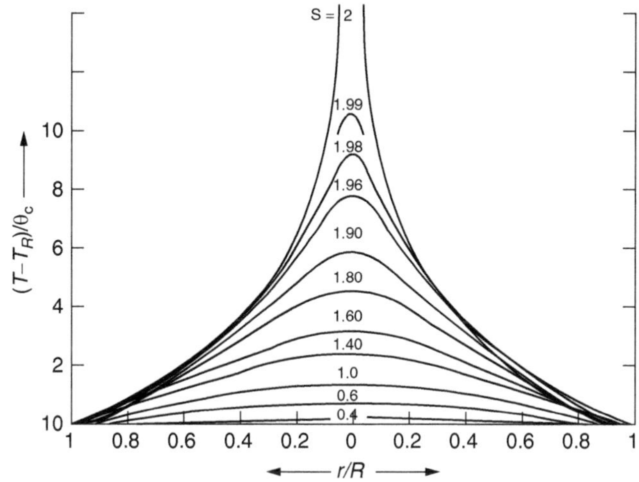

Fig. 6.7. Normalized temperature distribution in an electrically loaded cylinder with the normalized electric power (s) as family parameter

Assuming, for simplicity, cylindrical symmetry with neutral electrodes at the top and bottom surfaces and heat dissipation through the lateral surfaces, using Newton's cooling law, one has:

$$-\kappa \frac{\mathrm{d}T}{\mathrm{d}r}\bigg|_{r0} = h(T_{r0} - T_0), \tag{6.4}$$

with κ the heat conductivity, h the surface heat transfer coefficient, r_0 the cylinder radius, and T_0 the ambient temperature. With $|\mathbf{grad}\,\psi| = F = v/d$, and for steady state, one obtains from (6.3)

$$\sigma F^2 + \frac{\kappa}{r}\frac{\mathrm{d}}{\mathrm{d}r}\left(r\frac{\mathrm{d}T}{\mathrm{d}r}\right) = 0. \tag{6.5}$$

When approximating[3] $\sigma(T)$ of (6.2) with

$$\sigma = \sigma_0 \exp\left(\frac{T - T_0}{\Theta_\mathrm{c}}\right) \qquad \text{with} \qquad \Theta_\mathrm{c} = \frac{kT^2}{E_\mathrm{c} - E_\mathrm{Fn}} \tag{6.6}$$

and assuming a temperature-independent thermal conductivity, one can integrate (6.5), obtaining

$$\sigma_0 F^2 \exp\left(\frac{T - T_0}{\Theta_\mathrm{c}}\right) = \frac{2\kappa\Theta_\mathrm{c}}{r^2}\frac{\alpha^2}{\cosh^2[\alpha(\ln r + \beta)]} \tag{6.7}$$

[3] This approximation is obtained by development of (6.2) around T_0.

with integration constants α and β that can be obtained from the symmetry condition

$$\kappa \frac{\mathrm{d}T}{\mathrm{d}r}\bigg|_{r=0} = 0, \qquad \text{yielding} \qquad \alpha^2 = 1 \tag{6.8}$$

and from the total energy balance[4]

$$\mathcal{N} = \int_{\mathcal{V}} \sigma F^2 \mathrm{d}\mathcal{V} = -\int_{\mathcal{O}} \kappa \frac{\mathrm{d}T}{\mathrm{d}r}\bigg|_{r_0} \mathrm{d}\mathcal{O}, \tag{6.9}$$

where $\mathcal{N} = IV$ (with I the current and V the applied voltage), $\mathcal{V}$ is the volume of the cylinder and $\mathcal{O}$ its surface. Equation (6.9) yields

$$IV = 2\pi r_0 \ell \kappa \frac{\mathrm{d}T}{\mathrm{d}r}\bigg|_{r0} = 4\pi \ell \kappa \Theta_c \Big\{1 + \tanh(\ln r_0 + \beta)\Big\}, \tag{6.10}$$

with ℓ the length of the cylinder and r_0 its radius. From (6.10) β can be obtained. For simplicity of the following expressions, β is replaced with

$$s = 1 + \tanh(\ln r_0 + \beta) = \frac{IV}{\ell} \cdot \frac{1}{4\pi\kappa\Theta_c} = \mathrm{const}\mathcal{N}, \tag{6.11}$$

which varies between $0 \leq s < 2$ (since tanh is limited between ± 1). This integration constant is proportional to the power density ($\mathcal{N}/\ell$) in the cylinder and inversely proportional to the thermal conductivity. s can be used as normalized electric power.

Inside the cylinder, the temperature distribution can therefore[5] (by introducing (6.8) and (6.11) into (6.7)) be expressed as

$$T(r) = T_{r_0} + 2\Theta_c \ln \left[\frac{1}{1 - \frac{s}{2}(1 - \frac{r^2}{r_0^2})}\right]. \tag{6.12}$$

The heat-flux density at any surface within the cylinder is given by

$$-\kappa \frac{\mathrm{d}T}{\mathrm{d}t} = \frac{2\kappa\Theta_c}{r_0^2} \frac{sr}{1 - \frac{s}{2}(1 - \frac{r^2}{r_0^2})}, \tag{6.13}$$

which increases monotonically with the electric power s. For $s < 1$ the heat flux is largest at the outer surface, i.e., at r_0. For $s > 1$ its maximum lies inside the cylinder, indicating an inflection point for $T(r)$. The position of this inflection point is given by

---

4 The electrical energy supplied to the cylinder volume must be equal to the thermal energy dissipated from the cylinder surface in steady state.

5 The temperature is now expressed as an excess temperature $T - T_{r0}$ over the surface temperature T_{r0}. The latter is given by Newton's law.

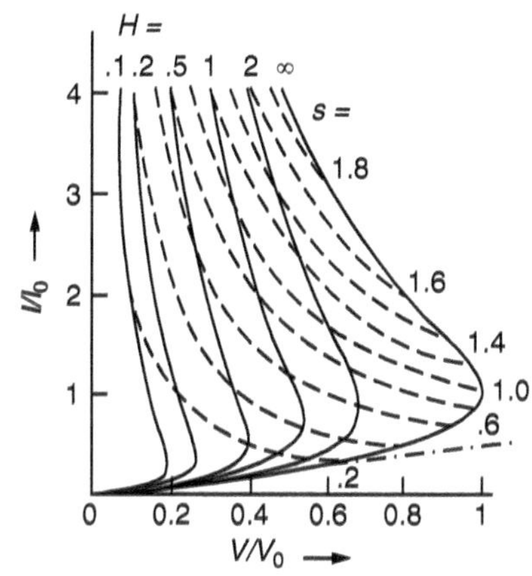

Fig. 6.8. Computed current–voltage characteristics (in normalized units) in the S-shaped region. Family parameter is the reduced heat transfer coefficient H. *Dashed curves* are equi-power curves with the normalized electric power as family parameter

$$r_i = r_0\sqrt{\frac{2-s}{s}}. \tag{6.14}$$

As we will see later (Fig. 6.8), this inflection point occurs only when the characteristic contains a negative differential resistance branch.

The surface temperature is given by

$$T_{r_0} = T_0 + \frac{IV}{2\pi\ell r_0 h}. \tag{6.15}$$

The excess temperature distribution in the cylinder, normalized to Θ_c is shown in Fig. 6.7 with the normalized electric power s as family parameter.

From (6.7), (6.11), and (6.4) one can express the current and the applied voltage as a parametric function of s:

$$I(s) = 2\pi\sqrt{2\kappa\Theta_c\sigma_0 r_0^2}\sqrt{\frac{s}{2-s}}\exp\left(\frac{s}{H}\right) = I_0\sqrt{\frac{s}{2-s}}\exp\left(\frac{s}{H}\right) \tag{6.16}$$

and

$$V(s) = \ell\sqrt{\frac{2\kappa\Theta_c}{\sigma_0 r_0^2}}\sqrt{s(2-s)}\exp\left(-\frac{s}{H}\right) = V_0\sqrt{s(2-s)}\exp\left(-\frac{s}{H}\right) \tag{6.17}$$

with $H = hr_0/\kappa$. From (6.16) and (6.17) one deduces the current–voltage characteristic. The electric resistance

$$R(s) = \frac{V}{I} = R_0\left(1-\frac{s}{2}\right)\exp\left(-\frac{2s}{H}\right) \quad \text{with } R_0 = \frac{\ell}{\pi\sigma_0 r_0^2} \tag{6.18}$$

is rapidly decreases with increasing electric power. Figure 6.8 shows a family of these normalized current–voltage characteristics with a normalized heat transfer coefficient as family parameter. All of these characteristics show for sufficient electric power s a non-monotonic behavior in V with a negative differential resistance branch:

$$\frac{\mathrm{d}V}{\mathrm{d}I} = R(s)\left\{\frac{2-s}{1+\dfrac{s(2-s)}{H}} - 1\right\} < 0 \tag{6.19}$$

as shown in Fig. 6.8. The maximum bias point, at which the differential resistance vanishes, defines the critical power

$$s_{\mathrm{c}} = 1 + \frac{H}{2} - \sqrt{1 + \frac{H^2}{4}}, \tag{6.20}$$

yielding a value $s_{\mathrm{c}} < 1$, which decreases with a decreasing heat transfer coefficient. When $H \gg 1$ (i.e., heat can be well-dissipated from the surface), $s_{\mathrm{c}} \simeq 1$ and the maximum bias point is given by

$$V_{\max} = V_0 \quad \text{with} \quad R(V_{\max}) = \frac{R_0}{2} \quad \text{for} \quad H \gg 1. \tag{6.21}$$

If, however, the heat transfer coefficient is very low ($H \ll 1$), the critical power is $s_{\mathrm{c}} = H/2$ and the maximum bias is given by

$$V_{\max} = V_0\sqrt{\frac{hr_0}{\kappa}} \quad \text{with} \quad R(V_{\max}) = R_0\exp(-1) \quad \text{for} \quad H \ll 1. \tag{6.22}$$

Here, the temperature profile remains relatively flat, while the surface temperature has risen substantially. A still larger fraction of the characteristic is transversed in the negative differential resistance branch.

The thermo-optical observation of the temperature distribution within current filaments in CdS platelets confirms the general behavior predicted by the theory (Fig. 6.9). A quantitative comparison, however, is hampered for reasons that the platelet geometry does not conform with the simplified cylindrical model.

The CdS platelet example given here is chosen for its easy observation of the typical current channel behavior. Similar properties exist in a wide variety of other semiconductors in the appropriate parameter range.

There is a large variety of kinetics in the channel formation and its movement through the crystal, including some pulsation that may be caused by a contribution of an N-type negative resistance regime.

Also the heating in the channel produces mechanical stress on the surrounding parts, that, if the high voltage is switched on fast it may lead to mechanical cracking and partial destruction of the current channel. All of this can be followed in the second part of the film on the Springer Web site.

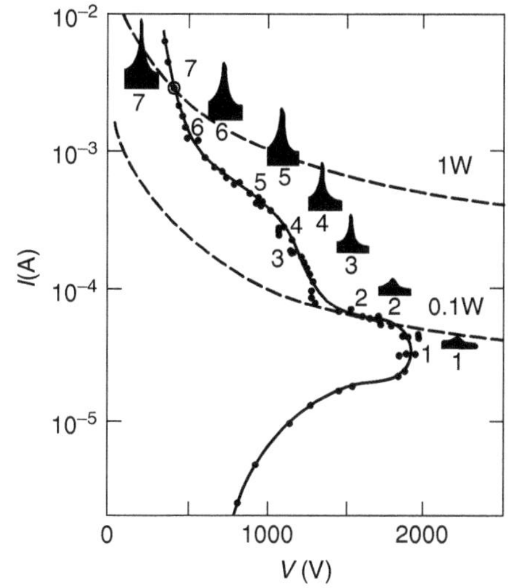

Fig. 6.9. Current–voltage characteristic of a CdS single crystal platelet in the range of S-shaped negative differential resistance, with temperature profiles (50° C/mm of the ordinate) of the current channel, measured approximately half way between the two electrodes; platelet width = 5 mm, thickness = 0.15 mm; the corresponding points at the current–voltage characteristic to each of the shown profiles are identified with numbers (After Böer et al. 1961)

6.2 General Discussion of Current Channels or Filaments

We will now proceed to a more general discussion of current channel formation that we, in more general terms may identify as current filaments. This presentation is based on the discussion by Schöll in Chap. 34 of Böer (2002).

First, we will consider the response of the system to small **space- and time-**dependent fluctuations of the homogeneous steady state. As we have seen in Chap. 5, under certain conditions, in particular when negative differential conductivity occurs, these fluctuations can grow, and lead to the bifurcation of spatial patterns. These may be modulated either in the direction of current flow or perpendicular to that direction. This implies an important distinction between domain-like and filamentary instabilities, respectively (Fig. 6.1). The general discussion of the filamentary instabilities will be the subject of this chapter. In general, a variety of coupled electromagnetic and transport modes exist, depending upon the different orientations of the electric-field fluctuation as well as the wave vector relative to the uniform electric field.

A detailed analysis of these linear modes and possible bifurcations has been performed for the case of generation–recombination instabilities (Schöll 1987), including in particular low-temperature impurity impact ionization breakdown. Here, we shall only give simple general arguments for the bifurcation of current filaments in case of S-shaped negative differential conductivity regimes, referred to below as *SNDC*. Although the analogy of field domains and current filaments, based upon the duality of N-shaped, here indicated as *NNDC* and *SNDC* characteristics, respectively, has been emphasized in the early investigations (Ridley 1963; Volkov and Kogan 1969; Bonch-Bruevich et al. 1975), there remain essential differences since the formation of field domains can already be understood effectively within a one-dimensional treatment, while an analysis of current filamentation requires the consideration of at least one transverse direction in addition to the longitudinal coordinate.

6.2.1 Filamentary Instability

In this section, we shall use a simple phenomenological model to demonstrate that *SNDC* gives rise to a filamentary instability. For a wide class of semiconductor systems the internal state can be characterized by a single variable a which corresponds to the internal degree of freedom relevant for the nonlinearity of the charge transport. The physical meaning of a might be electron temperature (Volkov and Kogan 1969) or concentration of excess carriers in bulk semiconductors (Schöll 1987), charge density stored in the quantum well of double-barrier resonant tunneling structures (Glavin et al. 1997; Mel'nikov and Podlivaev 1998; Meixner et al. 2000), interface charge density in a heterostructure hot electron diode (Wacker and Schöll 1994) or the voltage across one of the pn-junctions in thyristors (Gorbatyuk and Rodin 1992; Meixner et al. 1998a, b), etc. The variable $a(y,z)$ together with the voltage drop V across the device determines the local current density $j(a,V)$. The spatial variables y, z denote the two coordinates perpendicular to the direction of the current flow, while the x-coordinate, i.e., the spatial dependence in the direction of the current flow, has been eliminated from the transport equations. The reduced dynamical description in terms of the parameter a can be derived from a full three-dimensional transport model, the Poisson equation and continuity equations via adiabatic elimination of fast variables as done for various semiconductor systems (see, for instance Schöll 2001). This equation takes the form of a nonlinear reaction–diffusion equation:

$$\frac{\partial a(y,z,t)}{\partial t} = f(a,V) + D\left(\frac{\partial^2 a}{\partial y^2} + \frac{\partial^2 a}{\partial z^2}\right), \tag{6.23}$$

where the function $f(a,V)$ is determined by the specific nonlinear transport properties; in case of *SNDC* it is a nonmonotonic function of a which has three zeros a_1, a_2, a_3 in a certain interval of voltages V. They correspond to the low-conductivity state, the intermediate state with negative differential conductivity, and the high-conductivity state of the S-shaped characteristic, respectively. D is an effective diffusion coefficient which describes the lateral coupling.

We restrict ourselves to a rectangular sample of size $L_y \times L_z$ and assume Neumann boundary conditions. Then the stability of the spatially uniform steady state (a^*, V^*) can be obtained by linearizing (6.23) around (a^*, V^*) for small space and time-dependent perturbations

$$\delta a(y, z, t) = a_{n,m}(t) \cos\left(\frac{n\pi}{L_y} y\right) \cos\left(\frac{m\pi}{L_z} z\right) \tag{6.24}$$

with wave vectors $k_{y,n} = n\pi/L_y$ and $k_{z,n} = n\pi/L_z$. For constant voltage, the amplitudes of these linear modes develop according to

$$\frac{\mathrm{d}a_{n,m}(t)}{\mathrm{d}t} = \left[\frac{\partial f}{\partial a}(a^*, V^*) - D\left(\frac{(n\pi)^2}{L_y^2} + \frac{(m\pi)^2}{L_z^2}\right)\right] a_{n,m}. \tag{6.25}$$

This leads to the dispersion relation

$$\lambda = \frac{\partial f}{\partial a} - D(k_{y,n}^2 + k_{z,m}^2). \tag{6.26}$$

A similar analysis holds for a cylindrical geometry where the linear modes are mth order Bessel functions of the radial coordinate $J_m(kr)\exp(im\phi)$, and the lowest mode $m = 0$ is cylindrically symmetric (Schöll 1987).

On branches of the current–voltage characteristic which are stable for voltage-controlled conditions we have $\frac{\partial f}{\partial a} \equiv f_{\mathrm{a}} < 0$, and all spatial fluctuations are damped out. The negative differential conductivity states correspond to $f_{\mathrm{a}} > 0$, and hence spatial fluctuations with small wave vectors $k < \sqrt{f_{\mathrm{a}}/D}$ grow in time. Thus, an instability occurs if at least one of the sample dimensions L_y, L_z is larger than $L_{\mathrm{crit}} = \sqrt{D\pi^2/f_{\mathrm{a}}}$, i.e., the $n = 1$ or $m = 1$ mode grows.

These inhomogeneous fluctuations can lead to the bifurcation of stable current filaments, especially if the operating point is stable against homogeneous fluctuations. More precisely, depending on the sample geometry, either plane current layers or cylindrical current filaments are formed. Such stable current filaments are characterized by the spatial coexistence of two locally stable homogeneous steady states at a given voltage: a high-conductivity state embedded in a low-conductivity state. Thus, unlike in case of the Gunn domains discussed in Chap. 4, current filament formation requires a bistable system.

6.2.2 Stationary Filaments

As we have shown in Sect. 6.2.1 by a linear mode analysis, the spatially homogeneous steady state of negative differential conductivity ($SNDC$) is unstable against transversal spatial fluctuations which lead to the bifurcation of current filaments.

Let us now discuss the fully developed stationary current filaments within a simple one-dimensional drift-diffusion model which neglects longitudinal spatial variations, but takes fully into account the nonlinearities. The electric field can then be decomposed into $\mathbf{F} = \mathbf{F}_{\|} + \mathbf{F}_{\perp}$, where $\|$ and $\perp$ denote components parallel and perpendicular to the current flow. It will turn out in the sequel of our discussion that the perpendicular internal electric field $F_{\perp}$ is generated by the space charges in the filament wall; it is necessary to stabilize the filament.

Let us choose the coordinate system such that the drift current density j and the drift field $F_{\|} = F_x$ are in the x-direction, and the internal space charge field $F_{\perp} = F_z$ is in the z-direction. The sample is considered to be sufficiently small in y-direction so that we can assume homogeneity in this direction.

The constitutive equations in the steady state are similar to the equations given in Chap. 4, except that now the spatial dependence of all relevant quantities is upon the transverse coordinate z: They are given by the drift-diffusion equation for the vanishing transverse current density

$$\frac{\partial n}{\partial z} = -\frac{enF_z}{kT} \tag{6.27}$$

and Poisson's equation

$$\frac{\partial F_z}{\partial z} = \frac{1}{\varepsilon\varepsilon_0}\varrho(n, F). \tag{6.28}$$

Here, ϱ is the charge density which contains the essential nonlinearity since it is given by the densities of free electrons and electrons bound at donors and acceptors, which themselves depend upon n via the reaction kinetics. In the case of *SNDC*, $\varrho(n, F)$ has three zeros n_1, n_2, n_3 in a range of fields $F_{\mathrm{h}} < F < F_{\mathrm{th}}$, corresponding to these. Examples of generation–recombination mechanisms which lead to such dependencies will be discussed later.

The solutions $n(z)$ and $F_z(z)$ can be conveniently discussed in the (n, F_z) phase plane, in exactly the same way as it was done for the field-of-direction analysis discussed in the previous chapters. We briefly summarize the discussion given in detail by Schöll (1987). The solutions include filamentary profiles which describe the spatial coexistence of the two stable homogeneous phases $n_l = n_1$ (low conductivity) and $n_h = n_3$ (high conductivity). The condition for the existence of such a kink-shaped profile $n(z)$ with the boundary conditions

$$n(-\infty) = n_l, \quad n(\infty) = n_h \tag{6.29}$$

(or, equivalently, n_l and n_h interchanged) with a thin planar interfacial layer (filament border) can be cast into the form of a simple geometrical construction under the assumption $|F_z| \ll |F_x|$: an **equal areas rule**.

Solving (6.27) for F_z, differentiating and substituting into (6.28) gives

$$\frac{\partial}{\partial z}\left(\frac{1}{n}\frac{\partial n}{\partial z}\right) + \frac{e\varrho(n, F_x)}{\varepsilon\varepsilon_0 kT} = 0. \tag{6.30}$$

Equation (6.30) can be multiplied by $\frac{1}{n}\frac{\partial n}{\partial z}$ and integrated over $\int_{-\infty}^{+\infty} \mathrm{d}z$ yielding

$$\frac{e}{\varepsilon\varepsilon_0 kT}\int_{n_h}^{n_l} \varrho(n, F_x)\frac{\mathrm{d}n}{n} = 0, \tag{6.31}$$

where the boundary conditions $n(-\infty) = n_l$, $n(\infty) = n_{\mathrm{h}}$ and $\frac{\partial n}{\partial z}(-\infty) = \frac{\partial n}{\partial z}(\infty) = 0$ have been used.

Condition (6.31) singles out a specific value of the electric field F_x: the *coexistence* field F_{co} with $F_{\mathrm{h}} < F_{\mathrm{co}} < F_{\mathrm{th}}$. In terms of (n, F_z) phase portraits, the coexistence solutions correspond to heteroclinic orbits (saddle-to-saddle separatrices), while for all other values of F_x only homoclinic orbits (closed saddle-to-saddle-loops) exist. Solutions corresponding to finite boundaries may also be discussed in terms of the phase portrait (field-of-direction) (Schöll 1987).

The coexistence condition (6.31) may be visualized as an equal-area rule which requires that the two hatched areas in Fig. 6.10 are equal. Conditions for *cylindrical* current filaments were also developed in the form of equal areas rules for both unipolar and ambipolar generation – recombination mechanisms (Schöll 1986).

The current–voltage characteristic expected from this simple model which neglects inhomogeneities in the x-direction (hence F_x is proportional to the sample voltage V) is given by

$$I = e\mu_{\mathrm{n}}F_x \int n(z)\mathrm{d}z. \tag{6.32}$$

It is schematically shown in Fig. 6.11.

The full and dashed lines correspond to homogeneous steady states with positive and negative differential conductivity, respectively. The dash-dotted

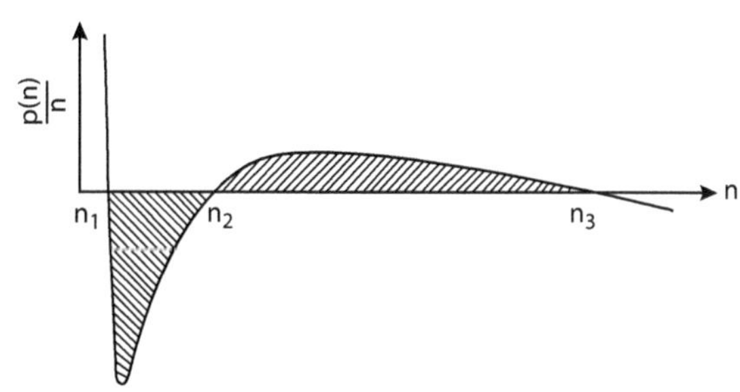

Fig. 6.10. The charge density $\varrho(n)/n$ is plotted vs. the carrier density n. The equal areas rule for coexistence of the low-conductivity phase n_1 and the high-conductivity phase n_3 with a planar interface is indicated by *hatched areas* (After Schöll 1982b)

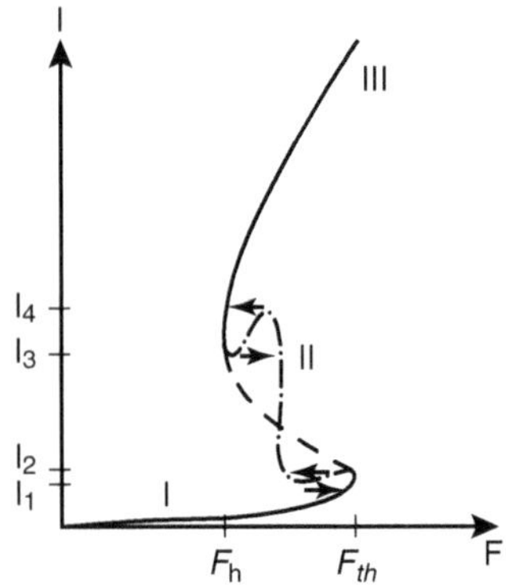

Fig. 6.11. Current–voltage characteristic including filamentary states during impurity impact ionization breakdown (in planar geometry). Stable and unstable homogeneous states are represented by *full* and *dashed lines*, respectively. Inhomogeneous filamentary states are shown as *dash-dotted lines*. Nonequilibrium phase transitions under current control between homogeneous states and current filaments are indicated by *arrows* (After Schöll 1987)

branch is associated with a current filament. Its almost vertical part corresponds to the lateral widening of the filament with strongly increasing current while the non-monotonic parts at its lower and upper ends describe the nucleation of a current filament from a homogeneous low-conductivity state (I), and its transition to a homogeneous high-conductivity state (III), respectively.

As discussed elsewhere (Schöll 1987; Schimansky-Geier et al. 1991), there exist profound analogies with nucleation phenomena, like metastability and supersaturation, also occurring in equilibrium phase transitions. Under approximately current-controlled conditions, the system switches along a horizontal load line between the different branches as indicated by arrows in Fig. 6.11. Note that there are small regions of NDC just before the switches occur, as observed experimentally (Kostial et al. 1995).

6.2.3 Spatiotemporal Dynamics of Filaments

The stationary current filaments may become unstable under certain conditions, resulting in oscillatory spatiotemporal dynamics. A menagery of complex spatiotemporal instabilities, such as breathing, spiking, rocking, or transversally traveling filaments, including various scenarios leading to spatiotemporal chaos, have been found in experiments and in computer simulations (Schöll et al. 1998; Schöll 2001). Some systematics can be brought into those phenomena by observing that the nonlinear spatiotemporal dynamics is governed by different competing spatial and temporal instabilities (Bose et al. 2000).

Generally, the semiconductor is governed by the interplay of the internal system dynamics, which is given by nonlinear transport equations such as, e.g., the reaction–diffusion system (6.23) or the drift-diffusion equations (6.27), and Kirchhoff's law for the external circuit the semiconductor system is operated in, which takes the form of a global constraint for the current density $j(y, z)$:

$$RC \frac{\mathrm{d}V}{\mathrm{d}t} = V_0 - V - R \int_G j(y, z)\mathrm{d}y\mathrm{d}z, \tag{6.33}$$

where R is the series load resistance, C is the capacitance of the external circuit, V_0 is the applied voltage, and G is a two-dimensional spatial domain in the plane perpendicular to the direction of the current flow. Models of this type are called activator–inhibitor systems since they typically involve one activating, i.e., destabilizing internal variable, and one inhibiting component, e.g., the voltage. They have been advanced for a wide class of semiconductor systems exhibiting both S- and Z-shaped bistability (see Schöll 2000, and references therein).

Depending upon the driving current or bias, the ratio of the timescales of the variables involved, and the lateral dimensions, a variety of different situations may arise. The negative differential conductivity state may experience a filamentary instability as well as a Hopf bifurcation leading to the onset of spatially uniform limit-cycle oscillations if the capacitance of the circuit is large enough. If both bifurcations coincide, a so-called codimension-two bifurcation arises. Such degenerate bifurcations have also been fond in other, *locally* coupled semiconductor systems (Heidemann et al. 1993; Wacker et al. 1995; Meixner et al. 1997a,b, 2000).

Complex spatiotemporal dynamics occurs if the spatially uniform steady state of negative differential conductivity simultaneously experiences a spatial instability leading to the formation of a stationary filament, and if this filament is unstable with respect to temporal oscillations (Alekseev et al. 1998), while at the same time the uniform steady state is *not* unstable with respect to uniform limit-cycle oscillations (Wacker and Schöll 1995). Then, a mixed spatiotemporal mode corresponding to large-amplitude relaxation-type oscillations of a current filament – *spatiotemporal spiking* – is possible. Chaotic spiking scenarios occur as well. Such behavior has been found numerically for one-dimensional (Wacker and Schöll 1994; Bose et al. 1994; Francheschini et al. 1999; Kehrt et al. 2009) as well as for two-dimensional (Bose et al. 2000; Stegemann and Schöll 2007) spatial domains, and confirmed by experiments on *pnpin* diodes (Niederostheide et al. 1996). This indicates that current or voltage oscillations may occur even if there is no oscillatory instability of the uniform state.

Another example is a double-barrier resonant tunneling diode (DBRT), which exhibits a Z-shaped (bistable) current–voltage characteristic (Schöll 2001; Goldmann et al. 1987) The DBRT is a semiconductor nanostructure which consists of one GaAs quantum well sandwiched between two AlGaAs barriers along the z-direction. The quantum well defines a two-dimensional electron gas in the x–y plane. We include the lateral redistribution of electrons in the quantum well plane giving rise to filamentary current flow (Meixner et al. 2000; Cheianov et al. 2000; Rodin and Schöll 2003). Complex chaotic scenarios including spatiotemporal breathing and spiking oscillations have been found in a simple deterministic reaction–diffusion model with one lateral dimension x (Schöll et al. 2002) as well as with two lateral dimensions x, y (Stegemann and Schöll 2007). The steady states are detemined by the intersection of the load line with the current–voltage characteristic. Besides up to three spatially homogeneous steady states, there is a spatially inhomogeneous filamentary state. The latter may become unstable in a supercritical Hopf bifurcation giving rise to spatiotemporal patterns in the form of breathing current filaments (Unkelbach et al. 2003). More complex, chaotic patterns are revealed in a period-doubling route to chaos. Such complex filamentary patterns can also be induced merely by noise, e.g., shot noise in the equation for the carrier density (Stegemann et al. 2005). The control of such unstable or irregular dynamics and the suppression of chaos is a major issue in applied nonlinear science (Schöll and Schuster 2008). Applying a time-delayed feedback loop in the operating circuit, unstable breathing or spiking patterns can indeed be stabilized both in the deterministic case (Baba et al. 2002; Beck et al. 2002; Unkelbach et al. 2003; Kehrt et al. 2009) and for noise-induced dynamics (Stegemann et al. 2006; Schöll et al. 2008; Majer and Schöll 2009).

The described scenarios of complex spatiotemporal behavior of current filaments are applicable to bistable systems whose internal state can be modeled by a *single* activator variable a and whose global inhibition is due to an external constraint. Another important mechanism for complex behavior occurs in systems with two mechanisms of inhibition acting on different time and space scales. The corresponding models are globally coupled two-component activator–inhibitor systems (Niederostheide et al. 1994, 1996, 1997), which are described by two internal local variables (activator and inhibitor) and one inhibiting global constraint and have local and global mechanisms of inhibition; and three-component activator–inhibitor models (Datsko 1997; Schenk et al. 1997; Gurevich et al. 2004; Rodin 2005) with three local variables and two local mechanisms of inhibition. Depending on the hierarchy of relaxation times and diffusion lengths, complex behavior in the form of traveling (Niederostheide et al. 1997; Schenk et al. 1997), breathing (Niederostheide et al. 1994), or spiking filaments (Datsko 1997) is found. These mechanisms of complex behavior are not related to degenerate bifurcations and can be expected to be more robust, but they are generally based on much more specific models and are therefore less universal.

The analysis of experimental data shows that degenerate bifurcations (Niederostheide et al. 1996) as well as multiple inhibition (Datsko 1997) can play a crucial role in complex dynamics, depending on the charge transport mechanism in a particular semiconductor system.

6.3 Examples of More Complex Filament Formation

The first direct observation of current filaments in the regime of *SNDC*, using an electro-thermo-optical method, has been reported by Böer et al. (1958, 1959) and Böer (1961), who also pointed out the similarity between current filamentation in semiconductors and plasma discharge.

Later on, current filaments have attracted interest in connection with *pin*-diodes (Barnett 1970; Jäger et al. 1986; Purwins et al. 1987; Symanczyk et al. 1991), amorphous chalcogenide films (Bosnell and Thomas 1972; Petersen and Adler 1976; Adler et al. 1980), low-temperature impurity breakdown in n-GaAs and p-Ge (Schöll 1982a, 1987; Aoki and Yamamoto 1983; Mayer et al. 1987; Brandl et al. 1989; Peinke et al. 1992; Aoki 2000a), and with *pnpin*-structures (Niedernostheide et al. 1992, 1993, and 1996) and thyristors (Gorbatyuk and Rodin 1990, 1992, 1997) at room temperature. The analogy between semiconductors and gas discharge has later been revived by investigations of pattern formation in gas discharge systems (Purwins et al. 1987) and by detailed simulations of the spatiotemporal dynamics of current filament formation (Gaa et al. 1996a; Schwarz et al. 2000; Murawski et al. 2005). Much theoretical and numerical work has been done following the seminal early papers by Böer (1961), Ridley (1963), Volkov and Kogan (1967), Kogan (1968), Böer and Döhler (1969), Bass et al. (1970), Varlamov and Osipov (1970), Kerner and Osipov (1976), for reviews see Schöll (1987, 2001), Shaw et al. (1992), Kerner and Osipov (1994), Schöll et al. (1998).

Current filaments have been directly observed by a number of spatially resolved sophisticated experimental techniques ranging from elaborate non-invasive techniques like the thermal shift of the optical absorption edge (Böer et al. 1959, 1961; Böer 1960, 1961, 1971), the quenched photoluminescence (Eberle et al. 1996; Aoki 1999) to invasive methods like scanning electron microscopy (Mayer et al. 1988; Wierschem et al. 1995), scanning laser microscopy (Brandl et al. 1989; Spangler et al. 1994; Kukuk et al. 1996), and potential probe measurements (Baumann et al. 1987; Niederostheim et al. 1992). We will now present another type of example, the impurity breakdown in GaAs.

6.3.1 Impurity Breakdown in n-GaAs

Another important example which has attracted much interest both experimentally and theoretically is associated with impact ionization breakdown of doped semiconductors at low temperatures.

Current filaments have been observed in the regime of low-temperature impurity breakdown in *n*-GaAs thin films and other materials where *SNDC* occurs (e.g., Brandl et al. 1987, 1989; Mayer et al. 1988; Eberle et al. 1996; Aoki and Fukui 1998). Being a nonlinear autocatalytic generation–recombination process, impurity impact ionization induces a nonequilibrium transition between a low-conductivity and a high-conductivity state and may lead to a menagery of spatiotemporal instabilities including self-generated oscillations and chaos as first observed by Aoki (1981) and Teitsworth et al. (1983), and reviewed, e.g., in Schöll (1987, 1992, 2001), Abe (1989), Peinke et al. (1992), Niederostheide (1995), Aoki (2000a).

If a critical threshold voltage is applied to the semiconductor sample, the free carrier concentration grows in an avalanche-like manner, and the current increases by orders of magnitude at an almost constant voltage due to current filamentation and lateral growth of filaments. Stabilization to avoid runaway is accomplished by including a series resistance of sufficient magnitude.

Current filaments in thin high-purity n-GaAs epitaxial layers have been visualized experimentally by a method which is based on the quenching of impurity and exciton luminescence under the influence of an electric field in the prebreakdown regime (Eberle et al. 1996). Inside a current filament, donor states are depleted due to impact ionization. This leads to a reduced rate of radiative transitions between donors and acceptors as well as exciton recombination. As a result, the spatial distribution of photoluminescence reproduces the density of free carriers in the semiconductor layer. Filaments appear as dark stripes in images recorded with an infrared sensitive camera.

Earlier theoretical work has been mostly restricted to one-dimensional treatments and simple phenomenological models (Schöll 1987). Periodically breathing filaments (Schöll and Drasdo 1990; Kunz and Schöll 1992) and the intermittent and chaotic behavior of laterally traveling filaments (Hüpper et al. 1993a,b) have been studied in one-dimensional simulations, where only the transverse spatial coordinate perpendicular to the current flow was taken into account.

The dipolar electric field between two point contacts was included in a phenomenological, effectively one-dimensional model for current filaments in n-GaAs (Novák 1995). Microscopic analyses of low-temperature impurity breakdown have been based upon Monte Carlo (MC) simulations (Kuhn et al. 1993; Quade et al. 1994) for p-Ge and for n-GaAs (Kehrer et al. 1995; Kostial et al. 1995). Using such microscopic models the nascence of current filaments has been simulated for two-dimensional simplified point-contact geometries (Gaa et al. 1996a,b; Schöll et al. 1997), including also the effect of a transverse magnetic field (Kunihiro et al. 1997).

More sophisticated modeling of current filamentation which includes the microscopic scattering and generation–recombination processes as well as realistic contact geometries on two-dimensional spatial domains have enabled a detailed quantitative comparison between theory and experiment for various sample and contact configurations (Schwarz et al. 2000a,b, 2002; Murawski et al. 2005).

In n-doped GaAs at temperatures of liquid helium, the carrier density in the conduction band, and hence the current density, is determined by the generation–recombination (GR) processes of carriers between the conduction band and the donor levels. The experimentally observed S-shaped current density-field relation in the regime of impurity breakdown can be explained if impact ionization from at least two impurity levels is taken into account (Schöll 1982a). The generation–recombination cycle starts with impact ionization from the ground state and is sustained at lower fields by impact ionization of the much less strongly bound excited state. The state of the system can be characterized by the spatial distribution of the carrier densities in the conduction band $n(\mathbf{r},t)$ as well as in the impurity ground state $n_1(\mathbf{r},t)$ and the excited state $n_2(\mathbf{r},t)$, where $\mathbf{r}$ is the position vector and t denotes time.

In the following, an approach is adopted which combines Monte Carlo (MC) simulations of the microscopic scattering processes and GR kinetics with time-dependent rate equations for the macroscopic spatiotemporal dynamics of the carrier densities occurring on a much slower time-scale than the microscopic dynamics (Schwarz 2000a).

The temporal evolution of n is governed by the continuity equation

$$\frac{\partial n}{\partial t} = \frac{1}{e}\nabla\cdot\mathbf{j} + \phi(n, n_1, n_2, |\mathbf{F}|), \tag{6.34}$$

where $\mathbf{F}$ is the local electric field within the sample. Within the drift-diffusion approximation the current density $\mathbf{j}$ can be expressed as

$$\mathbf{j} = e(n\mu_n\mathbf{F} + D\nabla n) \tag{6.35}$$

with the field-dependent mobility μ_n and diffusion coefficient D. We assume the validity of the Einstein relation $D = \mu_\mathrm{n} kT/e$.

The rate ϕ of GR processes depends on the local values of the carrier densities in the conduction band and at the impurities, given by n, n_1, and n_2, respectively, and the strength of the electric field $F = |\mathbf{F}|$. Analogously, rates ϕ_1, ϕ_2 determining the temporal evolution of n_1, n_2 can be defined as

$$\frac{\partial n_i}{\partial t} = \phi_i(n, n_1, n_2, F) \tag{6.36}$$

with $i = 1, 2$.

In an explicit two-level model (Schöll 1982a), e.g., the GR rates are given by

$$\begin{aligned} \phi &= X_1^S n_2 - T_1^S n p_\mathrm{d} + X_1 n n_1 + X_1^* n n_2, \\ \phi_1 &= T^* n_2 - X^* n_1 - X_1 n n_1, \\ \phi_2 &= -\phi - \phi_1, \end{aligned} \tag{6.37}$$

where $p_\mathrm{d} = N_\mathrm{D} - n_1 - n_2$ is the density of ionized donors, N_D is the total density of donors, X_1^S is the thermal ionization coefficient of the excited level, T_1^S is its capture coefficient, X_1, X_1^* are the impact ionization coefficients from the ground and excited level, respectively, X^*, T^* denote the transition coefficients from the ground level to the excited level and vice versa, respectively.

The electric field is coupled to the carrier densities via Gauss' law

$$\varepsilon\varepsilon_0\nabla\cdot\mathbf{F} = e(N_\mathrm{D}^* - n_1 - n_2 - n), \tag{6.38}$$

where $N_D^* \equiv N_D - N_A$ holds with the compensating acceptor concentration N_A. From (6.34), (6.36)–(6.38) the charge conservation equation can be derived, given by $\nabla \cdot \mathbf{J} = 0$ with $\mathbf{J} = \varepsilon\varepsilon_0\dot{\mathbf{F}} + \mathbf{j}$, where $\mathbf{J}$ is the total current density composed of displacement current and conduction current densities.

The essential nonlinearities of the constitutive model equations (6.34)–(6.38) in the regime of low-temperature impurity breakdown are contained in the dependence of the GR coefficients upon n, n_1, n_2, and F. To derive these from a microscopic theory single particle MC simulations for a spatially homogeneous steady state have been performed (Kehrer et al. 1995). Hereby the S-shaped homogeneous current–voltage characteristic has been successfully reproduced. The macroscopic equations (6.34)–(6.38) were then solved on a two-dimensional spatial domain by an implicit finite element algorithm (Kunz et al. 1996) using fitted analytical representations of the MC data.

Figure 6.12 shows the current density distributions corresponding to five different operating points (a)–(e) on the current–voltage characteristic. They represent stationary current filaments for fixed V_0 and five values of the load resistance R increasing from (a) to (e). At $R = 100\,\Omega$ (a) the filament is very wide and plum-shaped (convex). With increasing R the operating point on the characteristic is shifted on the upper branch toward lower sample voltage and current, and the width of the filament decreases. Finally, at $R = 10\,\mathrm{M}\Omega$ (e) the operating point is located on the intermediate branch with negative differential conductance, and the filament becomes unstable and thin, and its current density drops by several orders of magnitude. Upon further increase of R (or likewise decrease of V_0 at fixed R), the filament fades away.

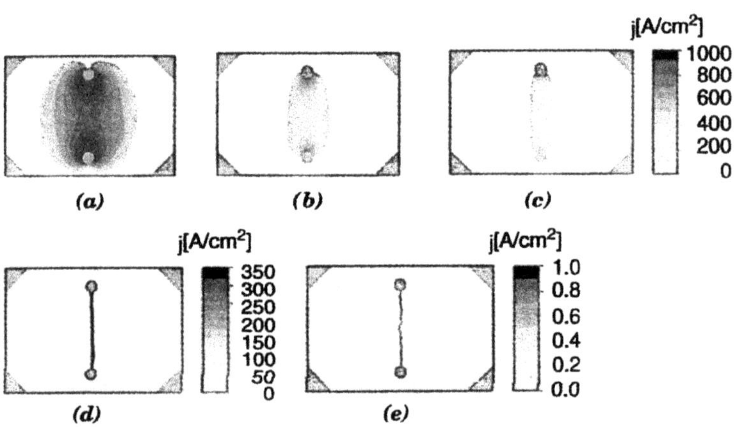

Fig. 6.12. Simulated current filaments in a rectangular sample ($5\,\mathrm{mm} \times 7.5\,\mathrm{mm}$) with two inner point contacts for an applied bias voltage $V_0 = 50\,\mathrm{V}$ and different load resistances R. The stationary current density $|\mathbf{j}(x,z)|$ is shown as a density plot. (**a**) $R = 100\,\Omega$, (**b**) $R = 500\,\Omega$, (**c**) $R = 1\,\mathrm{k}\Omega$, (**d**) $R = 10\,\mathrm{k}\Omega$, (**e**) $R = 10\,\mathrm{M}\Omega$ (After Schwarz et al. 2000a)

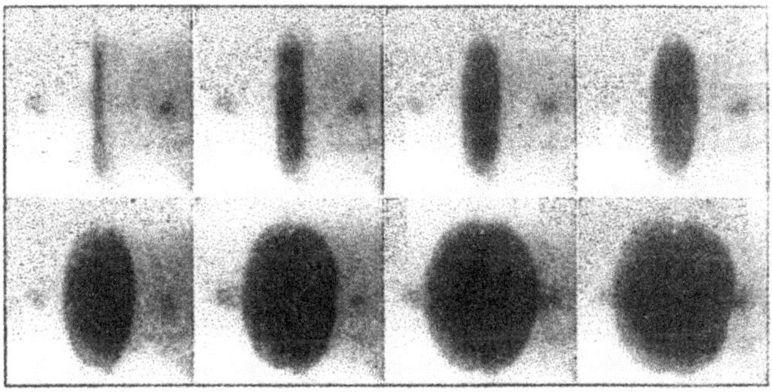

Fig. 6.13. Quenched photoluminescence images of current filaments in an n-GaAs sample with point contacts at 1.8 K with current increasing from (**a**) to (**h**): $I = 0.40$, 1.11, 1.57, 2.05, 3.42, 5.29, 6.70, 8.40 mA (After Schwarz 2000a)

The change of the shape and the width of the filament describes a transition from a pattern reflecting the dipolar electric field between the contacts (a) to a self-organized straight filament with parallel borders and constant width (d). The curving of the filament boundaries around the circular contacts with increasing current levels can be understood as a geometric effect within a reduced model (Novák et al. 1998).

In Fig. 6.13, experimental luminescence images comparable to these simulations are shown in the regime of filamentary current flow for various currents. The shape of the filaments and their change with rising current is in good agreement with the calculations. In spite of the inhomogeneous field distribution imposed by the point contacts, the filaments are formed as stripe-like structures with clearly defined borders. The borders remain parallel at lower currents and are continuously transformed to a convex shape at higher current levels.

If an additional magnetic field applied perpendicular to the epitaxial layer, this can sensitively affect the spatiotemporal instabilities in the regime of impurity impact ionization (Brandl 1987, 1990; Rau et al. 1991; Hirsch et al. 1994; Spangler et al. 1994). Stationary current filament configurations can either be destabilized resulting in nonlinear oscillations, or the filament patterns are deformed preserving their stability. Again the experimental results agree well with simulations (Schwarz et al. 2000a).

Samples with circular symmetry, i.e., *Corbino disks*, and concentric circular contacts have been intensely investigated. Such geometries offer the possibility to study pattern formation in samples without lateral boundaries, and are therefore of considerable interest. Spontaneous symmetry breaking of the current density distribution and various self-organized patterns of single and multiple current filaments have been observed experimentally in such samples

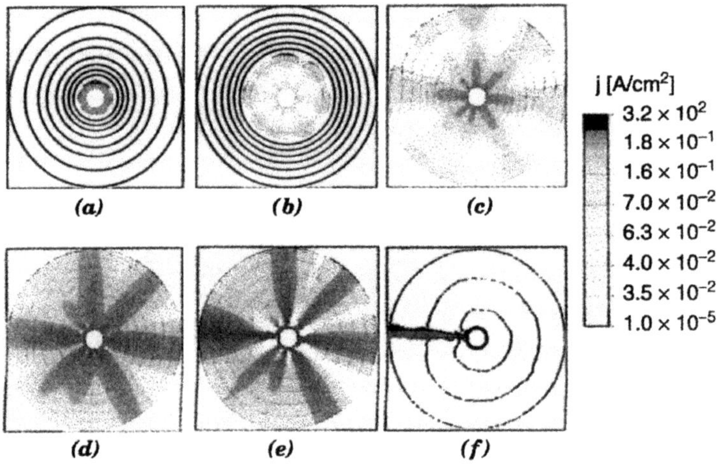

Fig. 6.14. Simulation of the formation of a current filament in a Corbino disk. The temporal evolution of the current density $|\mathbf{j}(x, y)|$ is shown for a circular sample of diameter 2.1 mm for a bias voltage $V_0 = 2\,\mathrm{V}$ and a load resistance $R = 10\,\mathrm{k\Omega}$. (**a**) $t = 0.2\,\mathrm{ns}$, (**b**) $t = 0.8\,\mathrm{ns}$, (**c**) $t = 2.3\,\mathrm{ns}$, (**d**) $t = 4.0\,\mathrm{ns}$ (**e**) $t = 8.0\,\mathrm{ns}$ (**f**) $t = 20.0\,\mathrm{ns}$ (After Schwarz 2000a)

(Hirschinger et al. 1997) and reproduced in simulations (Schwarz and Schöll 1997, Schwarz et al. 2000b, 2002).

Figure 6.14 shows an example of a simulation. A constant bias voltage $V_0 = 2\,\mathrm{V}$ is applied to the Corbino disk via a load resistor $R = 10\,\mathrm{k\Omega}$. Initially, the sample is in the insulating state. Subsequently, impact ionization multiplies the electron concentration at the injecting inner point contact and establishes a radially symmetric front that moves toward the circular anode (Fig. 6.14a). At a certain radius the circular front breaks up into several streamers (b), each of which, upon reaching the anode, forms a rudimentary radial current filament (c). The symmetry of this pre-filamentary state, i.e., the number of rudimentary filaments, depends on sample parameters as well as on the applied voltage. The pre-filaments then start growing in current while retaining approximately their original width (d). This leads to a rising total current I and thus, via the external load resistance, which acts as a global constraint due to Kirchhoff's law $V_0 = V + RI$, to a reduction in the sample voltage V. The pre-filaments therefore enter into a competition as a result of which only the filament which has first reached the outer contact survives (e). The other filaments slowly decay into the low-conducting state and vanish (f).

This winner-takes-all dynamics is characteristic of a global coupling as imposed by the load resistance, and is familiar from general nonlinear self-organizing processes under global constraints (Schimanski-Geier et al. 1991).

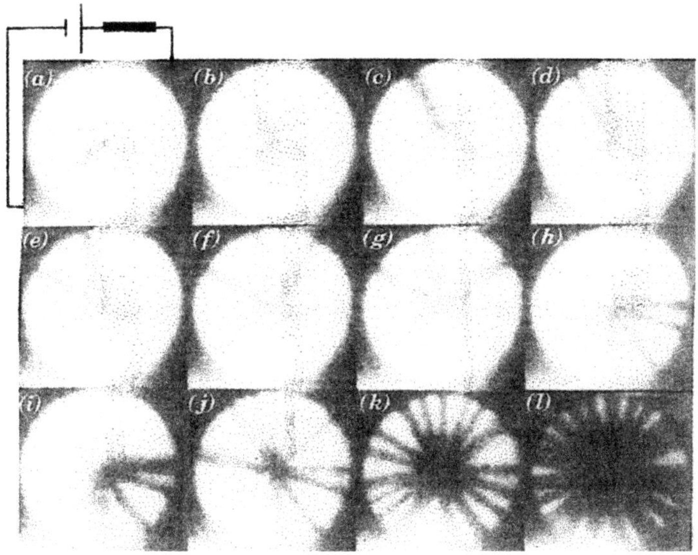

Fig. 6.15. Quenched photoluminescence images of filaments in an n-GaAs Corbino sample at 1.8 K with current increasing from (**a**) to (**l**): $I = 0$, 0.10, 0.43, 0.45, 0.47, 0.58, 0.77, 0.80, 1.54, 1.71, 4.08, 7.08 mA (After Schwarz et al. 2000a)

The experimentally observed stationary luminescence images of a Corbino sample are shown in Fig. 6.15. The figure shows the formation of multi-filamentary patterns upon increase of the current along a slowly varying ramp from zero up to 7 mA corresponding to a sample voltage of 3 V. At a critical value of the dc bias, the originally uniform state bifurcates into a single current filament being formed at a random position (b), in good agreement with Fig. 6.15f. Upon further increase of the current the filament approximately doubles its width (c) and a second bifurcation from one filament to two filaments occurs (d). The subsequent generation of additional filaments upon increasing dc bias occurs in the same way. With increasing current, all filaments grow in width and in most cases suddenly one of the filaments splits resulting in two filaments of smaller width. The angular distribution of the filaments becomes more and more symmetric with increasing number of filaments (k, l). The coexistence of multiple filaments can also be found within the numerical simulation, if the operating point is shifted to larger currents (Schwarz 2000b).

6.4 Summary and Emphasis

The creation of current channels or filaments occurs in systems with an S-shaped current–voltage characteristic. Stationary current channels that can be stabilized with a sufficiently large series resistor, as well as a menagery of spatiotemporal instabilities resulting in breathing, spiking, rocking, or laterally traveling filaments, and chaotic patterns, have been found.

Current filaments may be visualized directly by methods using the band-edge shift with increasing temperature; scanning electron or laser microscopy; potential probe techniques; or quenched photoluminescence.

Current filament formation is usually a pre-breakdown phenomenon. It can, however, be used for switching within the S-shaped characteristics of devices that are operated under appropriate external circuit conditions.

A

Educational Films

High-Field Domain and Current Channel Kinetics

A.1 High-Field Domain Analysis

In semiconductors, most interesting effects are of nonlinear form and caused by field and current inhomogeneities. Only in their rudimentary form these are dealt with by physical principals and their application to semiconductors. These may be the excitation and recombination processes of elementary particles such as electrons, photons, or phonons, space charges created by redistribution of carriers over defect centers, influenced by boundaries such as inhomogeneous doping, or electrodes. Others are caused by inhomogeneous heating, magnetic and electric fields, optical excitation and their mutual interaction. The result of these is a host of most interesting devices, from simple diodes to transistors to solar cells to luminescent diodes and semiconducting lasers and much more. And in modern technology, many of these devices are miniaturized, with millions of transistors and diodes in only a cubic millimeter, or others, like thin film solar cells may extend over many square meters, but the actual device performance is dictated by a minuscule sheet of only a few angstroms extending over the entire sheet. Most of what is causing all of the performance enhancement, only a few of us are now beginning to understand. While the technicians at the bench have managed, often by trial and error to optimize such devices to impressive performance.

A.1.1 Electro-Optical Tools

As often in life, the best way to see things that otherwise can only be anticipated is to "look into it," literarily, like the doctor looks under certain circumstances into a patient with X-rays, there is a way to look into a semiconductor, "literarily" with electro-optical effects. These are the tools which we found almost six decades ago that let us **see** the actual distribution of fields and currents inside a semiconductors, verify, what we theoretically predicted, and finding many new effects that we would never have come across with our previous analytical tools that were simply too course or too slow. This book describes in great detail the connection between what was observed with these new electro-optical tools and the theory that then became often more transparent.

But these descriptions, even in the best of books, necessarily are lacking of the dramatic experience when one sees the dynamics of developing such field and current inhomogeneities, and their kinetic behavior under the influence of changing external parameters. This was captured in the film.

Though the film was taken only 3 years after we developed first these electro-optical tools, it presents a widely diversified group of effects that are still the basics for today's understanding. Much of what was developed at that time was forgotten in the meantime or never seen since it was not felt relevant for the devices in development. It was not directly "applicable." So it seemed. The reason to show this film now is that we believe otherwise and we will say why.

A.1.2 Model Semiconductor Cadmium Sulfide

Like in industry, models are often essential to help us designing better machines or merchandize. An example is the automobile for which we use models in a wind tunnel to find out the best aerodynamic form to simultaneously provide the least air resistance while giving the most pleasing esthetical appearance and comfort for passengers and space for the engine and luggage. And so are models used in a large variety in almost all parts of the industry to improve design and make products more competitive. The originals are either too expensive to experiment with or not practical to use.

This is not much different for semiconductor devices. Most of our models are now produced inside of computers, they are the theoretical models. But as these mathematical models are enormously helpful, they are often misleading, as they have to be fed with a number of adjustable parameters. And here begin the dangers: since the parameters are adjusted so that the outcome of the mathematical model fits the experimental results, one needs to ask whether such parameter combination is unique and whether the model is sufficient to indeed explain the experiment or just simulates a result that within experimental errors copies what is found in nature. This can and has caused extremely costly development mistakes in the semiconductor or photocell industry, where in the latter much effort was wasted to chase down imagined series resistances that were derived from a simple computer model of a network of resistors and diodes and comparing these with the experimental results of current–voltage characteristics. The reasons were laying much deeper in the physics of the space charge behavior causing field inhomogeneities that were simply unexpected and could only be influenced by changing the doping distribution in those solar cells.

Some of the more sophisticated problems have only recently surfaced. For instance, when it became clear that a thin layer of CdS between the electrode and a thin film p-type semiconductor could enhance significantly the solar cell performance Böer 2009. Here, again a simple model of an adjustment of the reduction of reflection losses by the CdS interlayer were insufficient and grossly misleading.

This brings up to a much more actual level, in how for the results obtained so much earlier could one now obtain a better understanding of the physics of some of the device performances. Here, the CdS is helpful for another reason: it could act as a model semiconductor for many other materials.

Cadmium sulfide has a number of advantages, it is easy to grow as crystals and also as thin single crystal platelets; it is easy to change its electrical conductivity as a photoconductor, and a number of other effects at high electrical loads that were described in the book and helped creating a large variety of field and current distributions that are shown in the film. The CdS also has its absorption edge in the visible part of the spectrum that makes direct observation of electro-optical effects easy. And it provides by changing its temperature a means to shift kinetic effects into the range where direct cinematographic registration became easy.

With this, CdS became an ideal model substance, where many effects could be studied directly at macroscopic crystal platelets, while at different doping conditions and crystal dimensions similar effects occur that can no longer be observed visually.

With all this said, we feel encouraged to show this first documentary that was also designed to teach early students some of the intricacies of electronic behavior in solids.

A.2 The Film

This is a copy of parts of a DEFA film that was created in 1961 at the Humboldt University in Berlin to explain the state of the art of analyzing the field and current distributions in cadmium sulfide, and the development of electro-optical methods to make these distributions directly visible. The film was reorganized in 2009 and explained with English voiceover.

A.2.1 Field Inhomogeneities

This fist part of the film describes the different methods to obtain the field distribution in semiconductors and it shows a large variety of kinetic effects as observed inside the CdS by using the Franz Keldysh effect.

A.2.2 Current Distributions

This second part of the film deals with the development and kinetics of current channels in the pre-breakdown range, made visible by the thermal shift of the

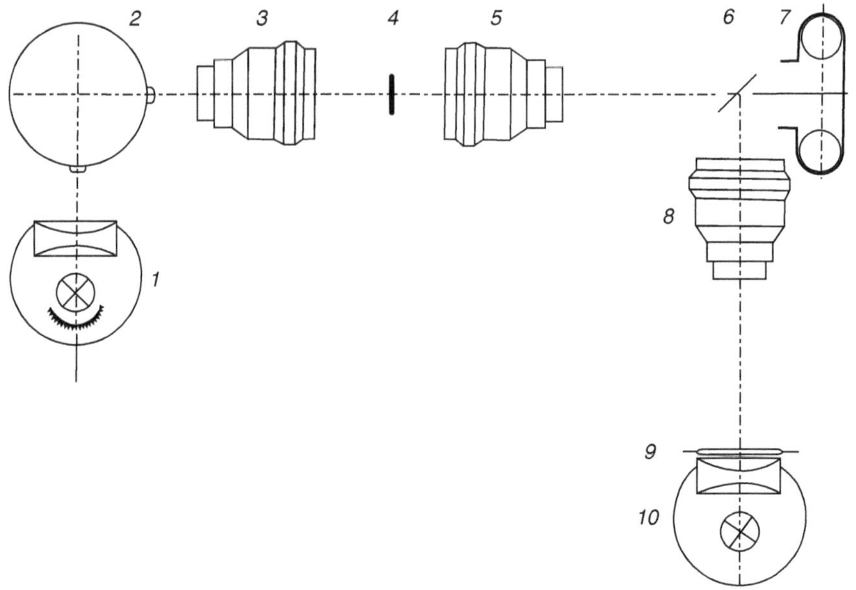

Fig. A.1. Experimental setup to film moving high-field domains and their internal kinetics. It also shows the blending-in of a projected volt and current meter at the bottom and top of some of the film scenes

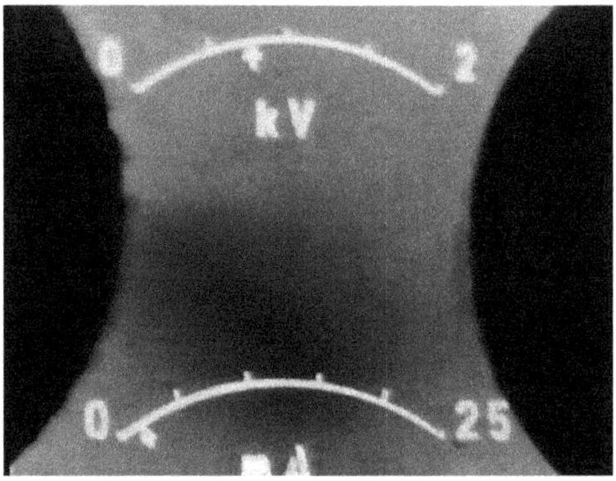

Fig. A.2. Frame from film with volt- and ammeter mirrored into t. Shown is a current channel as darkened part between the two electrodes

optical absorption edge. It provides a large sample of evidences to explain the physical processes occurring before thermal breakdown.

Fig. A.3. Celebration of the completion of the film with most of the involved people present. (Left Prof. K.W. Boër, middle Prof. right D. Palme)

Acknowledgments

Normally at the end of films, all important people are mentioned. Since this is not the case in conventional educational films, I will take this opportunity to thank a large number of professionals who were involved in making this film, from my colleagues, listed in the fore span, Drs. U. Kümmel, H.-J Hänsch, H. Lange, W. Wilhelm, and D. Palme, to the many not listed technicians and professionals at the DEFA film studio, helping in the mechanics of the film set-up, the animation, the tone, cutting and assembling of this film.

The film shown here with English voiceover is taken from parts of the original films that received first awards at an International Educational Film Festival in 1961 and 1962. The original films can be viewed at IWF-Wissen und Medien GmbH, D-37075 Göttingen, Nonnensteig 72 Tel: 49-551 5024-260.

References

Abraham E, Anderson PW, Licciardello DC, Ranakrishnan TV (1979) Phys Rev Lett 42:673

Adler D et al (1980) Handbook of semiconductors, vol 1. North-Holland, Amsterdam, p 805

Albrecht S, Reining L, Sole RD, Onida G (1998) Phys Rev Lett 80:4510

Amann A, Wacker A, Bonilla LL, Schöll E (2001) Phys Rev E 63:066207

Amann A, Peters K, Parlitz U, Wacker A, Schöll E (2003) Phys Rev Lett 91:066601

Amann A, Schöll E (2005) J Stat Phys 119:1069

Amann A, Schöll E (2005) Phys Rev B 72:165319

Andersen OK (1973) Solid State Commun 13:133–136

Anderson PW (1958) Phys Rev 109:1492

Ao R, Kuemmerl L, Haarer D (1993) Jpn J Appl Phys 32:5248

Aoki K, Yamamoto K (1983) Phys Lett A98:72

Aoki K, Yamamoto K, Mugibayashi N, Schöll E (1989) Solid State Electron 32:1149

Aoki K, Fukui S (1998) J Phys Soc Jpn 67:1106

Aoki K (2000) Solid State Commun 117:21

Aoki K (2000) Nonlinear dynamics and chaos in semiconductors, Institute of Physics Publishing, Bristol

Aulbur WG, Jonsson L, Wilkins JW (2000) Solid state physics, vol 54. Academic, New York

Aydinli A, Gasanly NM, Yilmaz I, Serpenguzel A (1999) Semiconductor science and technology, vol 14, p 657

Baba N, Amann A, Schöll E, Just W (2002) Phys Rev Lett 89:074101

Balkan N, Ridley B (1989) Superlattices Microstruct 5:539

Banghard EK (1988) In: Proceedings of the 20th IEEE Photovoltaic Specialists Conference, Las Vegas, p 717

Baumann H, Symanczyk R, Radchaus C, Pulwins II, Jäger D (1957) Phys Lett A123:421

Baylac B, Marie X, Amand T, Brusseau M, Barrau J, Shekun Y (1995) Surf Sci 326:161

Bergman MJ, Teitsworth SW, Bonilla LL, Cantalapiera IR (1996) Phys Rev B53:1327

Bimberg D, Grundmann M, Ledenstov NN (1999) Quantum dot heterostructures, Wiley, UK
Böer KW, Borchard W (1954) Z Phys Chem 203:145
Böer KW, Hänsch HJ, Kümmel U (1958) Naturwissenschaften 45:460
Böer KW, Hänsch HJ, Kümmel U (1959) Z Phys 155:170
Böer KW (1959) Z Physik 156:184
Böer KW (1960) Festkörperprobleme, Friedr. Vieweg & Sohn, Braunschweig, vol 1, p 38
Böer KW (1961) J Chem Solids 22:123
Böer KW, Hänsch HJ, Kümmel U, Lange H, Nebauer E (1961) Phys Stat Sol 1:19
Böer KW, Wilhelm W (1963) Phys Stat Sol 3:1704
Böer KW, Wilhelm W (1964) Phys Stat Sol 4:237
Böer KW (1964) Res Film 5:47
Böer KW (1965) Phys Rev 139 A:1949
Böer KW, Dussel GA (1967) Phys Rev 154:292
Böer KW, Bogus K (1968) Phys Rev 176:899
Böer KW, Ward JJ (1967) Phys Rev 154:757
Böer KW, Voss P (1968) Phys Rev 171:899
Böer KW, Döhler G, Dussel GA, Voss P (1968) Phys Rev 169, 700, 170:703
Böer KW, Döhler G (1969) Phys Stae Sol 36:679; Phys Rev 186:793
Böer KW, Dussel GA, Voss P (1969) Phys Rev 179:703
Böer KW, Dussel GA (1970) Phys Stat Sol 30, 375, 391
Böer KW, Ovshinsky SR (1970) J Appl Phys 41:2675
Böer KW (1971) Phys Stat Sol A 4:571
Böer KW, Stirn RJ, Dussel GA (1973) Phys Rev B 1433, 1443
Böer KW (1976) Phys Rev B 13:5373
Böer KW (1976) Phys Stat Sol (a) 49:13
Böer KW (1977) Phys Stat Sol (a) 40:355
Böer KS (1979) J Appl Phys 50:5356
Böer KW (1981) J Appl Phys 51:4518
Böer KW (1985) Ann Physik 42:371
Böer KW (1990) Survey of semiconductor physics, vol I, 1st edn. Van Nostrand Reihold, New York
Böer KW (2001) Phys Stat Sol 184:201
Böer KW (2000) Survey of semiconductor physics, vol II, 2nd edn. Wiley, New York
Böer KW, Gunnar T (2004) J Solar Energy 6:412
Böer KW (2009) Phys Stat Sol A 206:2665
Bonnet D (ed.) (1992) Thin Film Solar Cells, Int J Solar Energy 12(1–4).
Bonch-Bruevich VL, Zvyagin IP, Mironov AG (1975) Domain electrical instabilities in semiconuctors. Consultants Bureau, New York
Bonilla LL, Cantalapiedra IR, Bergmann MJ, Teitsworth SWS (1994) Semic Sci Technol 9:599
Bonilla LL, Galan J, Cuesta JA, Martinez FC, Molera JM (1994) Phys Rev 50:8644
Bonilla LL, Bulaschenko OM, Gerlán J, Kindellan M, Moscoso M (1996) Sol State Elect 40:161
Bonilla LL, Cantalapiera IR, Gomila G, Rubi JM (1997) Phys Rev E 56:1500
Bonilla LL, Hernando PJ, Herero MA, Kindelan M, Velazquez JJL (1997) Physica B Amsterdam 108:168
Bonilla LL, Kindelan M, Hernando PJ (1998) Phys Rev B58:7046

Borse PH, Deshmuk N, Shinde RF, Date SK, Kulkarni SK (1999) J Mater Sci 34:6087
Borisov VE, Sablikov VA, Chmil AI, Borisova IV (1999) Semiconductors 33:60
Bose S, Wacker A, Schöll E (1994) Phys Lett A195:144
Bose S, Rodin P, Schöll E (2000) Phys Rev E62:1778
Bosnell JR, Thomas CB (1972) Solid State Electron 15:1261
Bouhelal A, Albert JP (1989) Solid State Commun 69:713
Bouhelal A, Albert JP (1993) Physica B: Condensed Matter 6:255
Bradley AL, Doran JP, Hegarty J, Stanley RP, Oesterle U, Houdre R, Ilegems M (2000) J Luminescence 85:261
Branchu S, Pailloux E, Garem H, Rabier J, Demenet JL (1999) Phys Stat Sol (A) 171:59
Brandl A, Geisel T, Prettl W (1987) Europhys Lett 3:401
Brandl A, Kröninger W, Prettl W, Obermeier G (1990) Phys Rev Lett 64:212
Brandl A, Völcker M, Prettl W (1989) Appl Phys Lett 55:238
Briggs D, Seah MP (1990) Practical surface analysis, vol 1, 2nd edn. Auger and x-ray Photoelectron Spectroscopy, Wiley, Chichester
Brochard S, Rabier J, Grilhe J (1998) EPJ Appl Phys 2:99
Broser I (1967) Physics and chemsitry of II–VI compounds. In: Aven M, Prener JS (eds), North-Holland, Amsterdam
Broser I, Herfort L, Kallmann H, Martius U (1948) Z Naturf 3a:6
Bube RH (1992) Electronic properties of semiconductors, Cambridge University Press, Cambridge, UK
Bulashenko OM, Bonilla LL (1995) Phys Rev B 52:7849
Bulaschenko OM, Garcia MC, Bonilla LL (1995) Phys Rev 52:7849; 53:10008
Bulaschenko OM, Luo KJ, Grahn HT, Ploog K, Bonilla LL (1999) Phys Rev B60:5694
Butcher PN (1965) Phys Lett 19:546
Büttiker M, Thomas H (1977) Phys Rev Lett 38:78
Büttiker M, Thomas H (1978) Solid-State Electron 21:95
Büttiker M, Thomas H (1979) Z Phys B 34:301
Büttiker M, Thomas H (1999) Z Physik B34:301
Cantalapiedra IR, Bonilla LL, Bergmann MJ, Teitsworth SW (1993) Phys Rev B48:12278
Cao JC, Lei XL (1999) Pys Rev B59:2199
Cardona M, Ley L (eds) Topics in applied physics, vol 26. Springer, Berlin
Chakraborty PK, Biswas JC (1997) J Appl Physics 82:3328
Cheianov V, Rodin P, Schöll E (2000) Phys Rev B 62:9966
Chemla DS (1999) Semiconductors and Semimetals, ch 3, vol 58. Academic, New York
Chen R (1969) J Appl Phys 40:570
Chen TP, Chen LJ, Huang TS, Guo YD (1992) Semiconductor Science and Technology 7:300
Choi KK, Levine BF, Malik RJ, Walker J, Bethea CG (1987) Phys Rev 35:4172
Clark RJH, Hester RE (eds) (1998) Spectroscopy for surface science. Wiley, Chichester, New York
Coleman PD, Freeman J, Morkoc H, Hess K, Streetman BG, Keever M (1982) Appl Phys Lett 40:493
Datsko B (1997) Semiconductors 31:146

Conwell EM (1967) High-field transport in semiconductors. Academic, New York
Davidov VYu, Subashiev AV, Cheng TS, Foxon CT, Goncharuk LN, Smirnov AN, Zolotareva RN (1998) Mater Sci Forum 264–268:1371
Davydov VYu, Subashiev AV, Cheng TS, Foxon CT, Goncharuk IN, Smirnov AN, Solotareva RV, Ludyn WV (1998) Material Science Forum 264:1371
Di Bartolo B (1984) Energy transfer processes in condensed matter. Plenum, New York
Dong J, Drabold DA (1998) Phys Rev Letters 80:1928
Döttling R, Schöll E (1992) Phys Rev B45:1935
Döttling R, Schöll E (1993) Physica D, Amsterdam 67:418
Döttling R, Schöll E (1994) Solid State Electron 37:685
Drabold DA (1996) In: Thorpe MF, Mitkova MI (eds) Amorphous insulators and semiconductors, NATO ASI Series, vol 23, p 405
Drabold DA, Nakhmanson S, Zhang X (2001) In: Thorpe MF, Tichy L (eds) Properties and applications of amorphous materials. Kluwer, Dordech, p 221
Dussel GA, Böer KW (1970) Phys Stat Sol 39:375, 391
Dussel GA, Böer KW, Stirn RJ (1973) Phys Rev 7:1443
Eberle W, Hirchinger J, Margull U, Prettl W, Novak V, Kostial K (1996) Appl Phys Lett 68:3329
Ebert W, Vescan A, Borst TH, Kohn E (1994) IEEE Electron Device Lett 15:289
Efros AL, Rodina AV (1989) Solid State Commun 72:645
Esaki W, Chang LL (1974) Phys Rev Lett 33:495
Esser AE, Runge R, Zimmermann H, Langbein W (2000) Phys Stat Sol A 178:489
Fahrenbruch AL, Bube RH (1983) Fundamentals of solar cells. Academic, New York
Feng S (1990) Scattering and localization of classical waves in random media. World Scientific, Singapore
Feng PK (1999) Surface Science 429:L469
Fock W (1927) Arch Electrotechn 19:71
Franceschetti A, Zunger A (1997) Phys Rev Letters 78:915
Franceschini G, Bose S, Schöll E (1999) Phys Rev E60:5426
Franz W (1958) Z Naturforsch 13A:484
Frenkel JI (1938) Phys Rev 54:647
Fricke Ch, Hetiz R, Lummer B, Kutzer V, Hoffmann A, Broser A, Taud W, Heuken M (1994) J Crystal Growth 138:815
Fritzsche H, Ovshinsky SR (1960) J Noncryst Solids 4:464
Fritzsche H, (1997) Materials Res Soc Symposium Proc 467:19
Gaa M, Kunz RE, Schöll E (1996) Phys Rev B53:15971
Garcia-Cristobal A, Cantarero A, Trallero-Giner C, Cardona M (1998) Physica B: Condensed Matter 263:809
Geisz JF et al (2008) Appl Phys Lett 93:123505
Glavin B, Kochelap V, Mitin V (1997) Phys Rev B56:13346
Goldman VJ, Tsui DC, Cunningham JE (1987) Phys Rev Lett 58:1256
Gorbatyuk AV, Rodin PB (1990) Sovj Tech Phys Lett 16:529
Gorbatyuk AV, Rodin PB (1992) Sol Stat Electron 35:1359
Gorbatyuk AV, Rodin PB (1997) Z Physik B104:145
Grahn HT, Haug RJ, Müller W, Ploog K (1991) Phys Rev Lett 67:1618
Grahn HT, von Klitzing K, Ploog K, Döhler G (1991) Phys Rev B43:12094
Grahn HT, Castrup J, Ploog K, Bonilla LL, Garlan J, Kindelan M, Moscoso M (1995) Jpn J Appl Phys 34:4526

Gray PE (1967) Physics of electronics and circuit models. Wiley, New York
Gray JL, Schwartz RJ, Lundstrom MS, Nasby RD (1982) Proceedings of the 16th IEEE Photovoltaic Specialists Conference, San Diego, CA, p 437
Gray JL, Schwartz RJ (1984) Proceedings of the 17th IEEE Photovoltaic Specialists Conference, Kissimee, FL, p 1297
Green MA (1982) Solar cells. Prentice-Hall, Englewood Cliffs, NJ
Green MA (1997) Proceedings of the 26th IEEE Photovoltaic Specialists Conference, Las Vegas, p 717
Green MA, Emery K, Hishikawa Y, Warta W (2009) Progr Photovolt: Res Appl 17:85
Grenzer J, Schomburg E, Lingott I, Ignatov AA, Renk KF, Pietsch U, Zeimer U, Melzer BJ, Ivanov S, Schaposchnikov S, Kop"ev PS, Pavel'ev DG, Koschurinov Y (1998) Semicon Sci Technol 13:733
Greulich-Weber S (1997) Phys Stat Sol (A) 162:95
Gribnikov ZS, Mel'nikov MI (1973) Sovj Phys Sol State 7:2364
Gross EF, Permogor SA, Reznitz AN, Usarov EN (1974) Sovj Physics Semiconductors USSR 7:844
Guissi S, Bindi R, Iacconi P, Jeambrun D, Lapraz D (1998) J Phys D: Appl Phys 31:137
Gunn JB (1963) Sol State Commun 1:88
Gunn JB (1964) IBM J Res Devel 8:141
Gurevich SV, Bödeker HU, Moskalenko AS, Liehr AW, Purwins H-G (2004) Physica D 199:115
Gutsche E, Lange H (1964) Proceedings of the VII International Conference on Semiconductor Physics, Paris, p 129
Kalt H (1996) Optical properties of III-V semiconductors. The influence of multivalley band structure. Springer, Berlin
Kastrup J, et al. (1995) Phys Rev B 52:13761
Kastrup J, et al. (1997) Phys Rev B 55:2476
Hadley HC, Tyagi RC, Voss P, Böer KW (1972) Phys Stat Sol (a) 11, 523 and K 145
Heidemann G, Bode M, Purwins H (1993) Phys Lett A977:225
Hamakawa Y (1999) Appl Surf Sci 142:215
Hofbeck K, et al. (1996) Phys Lett A 218:349
Schöll E, Schuster HG (eds) (2008) Handbook of chaos control. Wiley-VCH, Weinheim, second completely revised and enlarged edition
Hao M, Sugahara T, Sato H, Morishima Y, Naoi Y, Romano LT, Sakai S (1998) Jpn J Appl Phys (2) 37:L291
Helm M, England P, Colas E, Derosa F, Allen Jr SJ (1989) Phys Rev Lett 63:74
Helgesen P, Finstad TG (1990) Proceedings of the 14th Nord Semiconductor Meeting U, Aarhus, p 323
Hendriks P, Zwaal EAE, DuBois JGA, Blom FF, Wolter JH (1991) J Appl Phys 69:302
Henisch HK (1984) Semiconductor contacts. Clarendon, Oxford, UK
Henry CH, Logan RA, Merrit FR (1978) J Appl Phys 49:3530
Hess K, Morkoc H, Shichijo H, Streetman BG (1979) Appl Phys Lett 35:469
Hilsum C (1962) Proc IRE 50:185
Higuera FJ, Bonilla LL (1992) Phys D, Amsterdam 57:161
Hirsch R, Kittel E, Flätgen G, Hübener RP, Parisi J (1994) Phys Lett A186:157

Hirschinger J, Eberle W, Prettl W, Niedernostheide FJ, Kostial H (1997) Phy Lett A236:249
Hizanidis J, Balanov AG, Amann A, Schöll E (2006) Int J Bifur Chaos 16:1701
Hizanidis J, Balanov AG, Amann A, Schöll E (2006) Phys Rev Lett 96:244104
Hizanidis J, Aust R, Schöll E (2008) Int J Bifur Chaos 18:1759
Hizanidis J, Aust R, Schöll E (2008) Phys Rev E 78:066205
Hopfbeck K, Grenzer J, Schomburg E, Ignatov AA, Renk KF, Pavel'ev DJ, Koschurinov Y, Melzer E, Ivanov S, Schaposchnilov S, Kop'ev PS (1996) Phys Lett A218:349
Hüpper G, Pyragas K, Schöll E Phys Rev B47:15515; B48:17633
Hybertson MS, Loui SG (1985) Phys Rev Lett 55:1418
Ibach H (1991) Electron energy loss spectrometer: The technology of high performance. Springer Verlag, Berlin
Jacobini C, Canali C, Ottoviani G, Cluaraula AA (1977) Sol State Electron 20:71
Jackson JD (1999) Classical electrodynamics, 2nd edn. Wiley, New York
Jäger D, Baumann H, Simanczyk R (1986) Phys Rev A117:141
Jappsen AK, Amann A, Wacker A, Schöll E, Schomburg E (2002) J Appl Phys 92:3137
Kahn AM, Mar DJ, Westervelt RM (1991) Phys Rev B43:9740
Kahn AM, Mar DJ, Westervelt RM (1992) Phys Rev Lett 46:369; Phys Rev B45:8342
Kamieniecki E (1990) Proc Electrochem Soc 90:1856
Kastrup J, et al. (1995) Phys Rev B 52:13761
Kastrup J, et al. (1997) Phys Rev B 55:2476
Kawamura Y, Wakita K, Asahi H, Kurumada K (1986) Jap J Appl Phys 25:L298
Keever M, Shichigo H, Hess K, Barnajee S, Witkovski L, Morkoc H (1981) Appl Phys Lett 38:36
Kehrer B, Quade W, Schöll E (1995) Phys Rev B51:7725
Kehrt M, Hövel P, Flunkert V, Dahlem MA, Rodin P, Schöll E (2009) Eur Phys J B 7:58
Keldysh LV (1958) Sov Phys JETP (Engl Transl) 4:665
Kerner BS, Osipov VV (1976) Sovj Phys 44:807
Kerner BS, Osipov VV (1994) Autosolitons. Kluwer, Dordrecht
Klemens PG (1966) Phys Rev 148:845
Knight BM, Peterson GA (1967) Phys Rev 155:393
Kobayashi Y, Kouzo K, Kamimuro H (1999) Sol State Commun 109:583
Kogan ShM (1968) Sovj Phys 27:656
Kohno M, Matsubara H, Okada H, Hirae S, Sakai T (1998) Jpn J Appl Phys (1) 37:5800
Kolev PV, Deen MJ, Kierstead J, Citterio M (1999) IEEE Trans Electron Devices 46:204
Kono S, Nagasawa N (1999) Sol State Commun 110:159
Korotov AN, Averin DV, Likharev KK (1993) Appl Phys Lett 62:3282
Kostial H, Asche M, Hey R, Ploog K, Kehra B, Quade W, Schöll E (1995) Semicond Sci Technol 10:775
Kroemer H (1964) Proc IEEE 52:1736
Kudlek GH, Pohl UW, Fricke Ch, Heitz R, Hoffmann A, Gutowski J, Broser I (1993) Physica B: Condensed Matter 185:325
Kuhn T, Hüpper G, Quade W, Rein A, Schöll E, Varani L, Reggiani L (1993) Phys Rev B48:1478

Kung P, Zhang X, Walker D, Saxler A, Piotrowski J, Rogalski A, Razeghi M (1995) Appl Phys Lett 67:3792
Kunihiro K, Gaa M, Schöll E (1997) Phys Rev B55:2207
Kunz RE, Schöll E (1992) Z Phys B89:289
Kunz RE, Schöll E, Gajewski H, Nürnberg R (1996) Sol State Electron 39:1155
Kukuk B, Reil F, Niedernorttheide FJ, Purwins HG (1996) Self-organization in activator-inhibitor systems, Wissenschaft und Technik Verlag, Berlin, p. 52
Kuo L-H, Salamanca-Riba L, Hofler GE, Wu B-J, Haase MA (1994) Proc SPIE 2228:144
Kwok SH, Grahn HT, Ramsteiner M, Ploog K, Prengl F, Wacker R, Schöll E, Murugkar S, Merlin R (1995) Phys Rev 51:8843
Lampert MA (1956) Phys Rev 903:1648
Larson P, Mahanti SD, Kanatzidis MG, (2000) Phys Rev B61:8162
Liou JJ, Wong WW, Juan JS (1990) Sol State Electron 33:845
Liu QZ, Lau SS (1998) Sol State Electron 42:677
Lueder H, Spenke E (1953) Z Techn Physik 16:53
Lundstrom MS, Schwartz RJ, Gray JL (1981) Sol State Electron 24:159
Luo KJ, Grahn HT, Ploog K, Bonilla lL (1998) Phys Rev Lett 81:1290
Milnes AG, Feucht DL (1972) Heterojunctions and metal–semiconductor junctions. Academic Press, New York
Maes W, De Meyer K, van Overstraeten R (1990) Sol State Electron 33:705
Magonov SN (1996) Surface analysis with STM and AFM: experimental and theoretical aspect of image analysis, VCH. Weinheim, New York
Majer N, Schöll E (2009) Phys Rev E 79:011109
Malpuech G, Kavokin A, Langbein W, Hvam JM (2000) Phys Rev Lett 85:650
Malyshkin V, McGurn AR, Leskova TA, Maradudin AA, Nieto-Vesperinas M (1997) Waves in Random Media 7:479
Marder MP (1998) Condensed Matter Physics. Wiley Interscience, New York
Mayer K, Gross R, Parisi J, Peinke J, Huebener R (1987) Sol State Commun 63:55
Mayer K, Parisi J, Huebener H (1988) Z Physik B71:171
McGurn AR (1999) Phys Lett A 251:322; Phys Lett A 260:314
McGurn AR (2000) Phys Rev B 61:13235
McKeever SWS (1985) Thermoluminescence of solids. Cambridge University Press, Cambridge
Meixner M, Bose S, Schöll E (1997) Physica D Amsterdam 109:928; Phys Rev B55:6690
Meixner M, Rodin P, Schöll E (1998) Phys Rev E58:2796; 5586
Meixner M, Rodin P, Schöll E, Wacker A (2000) Eur Phys J B13:157
Mel'nikov D, Podlivaev A (1998) Semicond 32:236
Miller D, Laikhtman B (1994) Phys Rev 50:18426
Mioc SL, Garland JW, Bennett BR (1996) Semicond Sci Technology 11:521
Mityagin YA, Murzin VN, Efimov YA, Rasulova GC (1997) Appl Phys Lett 70:3008
Mosekilde E, Feldberg R, Knudsen C, Hindsolm M (1990) Phys Rev 41:2298
Mosekilde E, Thomson JS, Knudson C, Feldbarg R (1993) Physica D Amsterdam 66:143
Morgenstern M, Wittneven C, Dombrowski R, Wiesendanger A (2000) Phys Rev Letter 84:5588
Mott NF, Gurney RW (1940) Electronic processes in ionic crystals. Oxford University Press, New York

Mott NF (1999) Metal–insulator transitions. Taylor and Francis, London
Mousseau N, Barkema GT (2000) Phys Rev B 40:1898
Murawski J, Schwarz G, Novák V, Prettl W, Schöll E (2005) J Appl Math Mech (ZAMM) 85:823
Nag BR (1980) Electron transport in compound semiconductors. Springer, Berlin
Nakamura K (1989) Phys Lett A138:396
Nataka H, Uddin A, Otsuka E (1992) Semiconductor Science and Technology 7:1266
Neudeck GW (1983) The pn-junction diode. Addison Wesley, Reading, MA
Neugroschel A, Sah CT (1996) Electron. Lett. 32:2280
Niedernostheide FJ, Arps M, Dohmen R, Willebrand K, Purwins HG (1992) Phys Stat Sol B 172:249; Phys Rev B46:7559
Niedernostheide FJ, Kreimer M, Schulze HJ, Purwins HG (1993) Phys Lett A180:113
Niedernostheide FJ, Artdes M, Or-Guil M, Purwins HG (1994) Phys Rev B49:7370
Niedernostheide FJ (1995) Non-linear dynamics and pattern formation in semiconductors and devices. Springer, Berlin
Niedernostheide FJ, Brillert C, Kukuk B, Purwins HG, Schulze HJ (1996) Phys Rev B54:14012; E54:1253
Niedernostheide FJ, Or-Guil M, Kleinkes M, Purwins HG (1997) Phys Rev E55:4107
Northrup D et al. (1964) Solid-State Electron 7:17
Nostrup DC, Thornton PR, Tresize KE (1964) Sol State Electron 7:17
Novak V, Wimmer C, Prettl W (1995) Phys Rev B52:9023
Nunoya N, Nakamura M, Yasamoto H, Tamura S, Arai S (2000) Jpn J Appl Phys 39:3410
Ovshinsky SR (1968) Phys Rev Lett 21:1450
Patra M, Schwarz G, Schöll E (1998) Phys Rev B57:1824
Peinke J, Parisi J, Rössler O, Stoop R (1992) Encounter with chaos. Springer, Berlin
Petersen KE, Adler D (1976) J Appl Phys 47:256
Piazza F, Christianen PCM, Maan JC (1977) Phys Rev B55:15591
Pines AS (1982) In: Cardona M (ed) Light scattering in solids. Springer, Berlin
Polezhaev VJ (1984) Crystals, growth, properties, and applications, vol 10. Springer, Berlin
Poole HH (1921) Philos Mag 42:488
Poole CP, Farach HA (1994) Data sources, vol 1. In: Handbook of electron spin resonance, Poole CP, Farach HA (eds), vol I. American Institute of Physics Press, New York
Prengel F, Wacker E, Schöll E (1994) Phys Rev 50:1705
Prince MB (1955) J Appl Phys 26:534
Pritchard J (1979) In: Roberts MW, Thomas JM (eds) Chemical physics of solids, vol 7. Chemical Society, p 166, New York
Purwins HG, Klempt G, Berkemeier J (1987) Festkörperprobleme 27:27
Quade W, Hüpper G, Schöll E, Kuhn T (1994) Phys Rev B49:13408
Racko J, Donoval D, Barus M, Nagel V, Grmanova A (1992) Sol State Electron 35:913
Rau U, Clauss W, Kittel A, Lehr M, Bayerbach M, Parisi J, Peinke J, Huebener R (1991) Phys Rev B43:2255
Redfield D, Bube R (1995) Mater Sci Symp Proc 325:335
Repins I, Contreras MA, Egaas B, DeHart C, Scharf J, Perkins CL, To B, Noufi R (2008) Prog Photovoltaics: Res Appl 16:235
Ridley BK (1961) Proc Phys Soc 78:293

Ridley BK, Wattkins TB (1961) Proc Phys Soc Lond 78:293
Ridley BK (1963) Proc Phys Soc 81:996
Ridley BK (1997) Electrons and phonons in semiconductor multilayers, Cambridge University Press, Cambridge, UK
Riviere JC, Myhra S (eds) (1998) Handbook of surface and interface analysis: methods of problem solving, Marcel Dekker, New York
Rockett A, Birkmire RW (1991) J Appl Phys 70:81
Rodin P, Schöll E (2003) J Appl Phys 93:6347
Rodin P (2005), Phys Rev B 71:085309
Rogozia M, Teitsworth SW, Grahn HT, Ploog KH (2002) Phys Rev B 65:205303
Rose A (1978) Concepts in photoconductivity and allied problems. Rober E. Krieger, Huntington, New York
Sah CT, Noyce RN, Shockley W (1957) Proc. IRE 45:1228
Sahyun MRV, Sharma DK, Serpone N (1995) J Imag Sci Technol 39:377
Sakurai T, Sakai A, Pickering HW (1989) Atom probe field ion microscopy and its applications. Academic, New York
Samuil VA (1995) Nonlinear dynamics and pattern formation in semiconductors and devices. Springer, Berlin
Sankin VI, Stolichnov IA (1994) JETP Lett 59:744
Schappe RS, Walker T, Anderson LW, Lin CC (1996) Phys Rev Lett 76:4328
Schenk CP, Or-Guill M, Bode M, Purwins HG (1997) Phys Rev Lett 78:3781
Schimansky L, Zülicke C, Schöll E (1991) Z Phys B84:433
Schlesner J, Amann A, Janson NB, Just W, Schöll E (2003) Phys Rev E 68:066208
Schottky W (1939) Z Phys 113:367
Schmidt KH, Linder N, Doehler GH, Grahn HT, Ploog K, Schneider H (1994) Phys Rev Letters 72:2769
Schmidt TM, Fazzio A, Caldas MJ (1995) Material Schience Forum 196:273
Schmitt OM, Banyai L, Haug H (2000) J Luminescence 87:168
Schmitt-Rink S, Chemla DS, Miller DAB (1989) Adv Phys 38:89
Schöll E (1982) Z Phys B46:23; B48:153
Schöll E (1986) Sol State Electron 29:687
Schöll E (1987) Non-equilibrium phase transitions in semiconductors. Springer, Berlin
Schöll E, Schwarz G, Wacker A (1998) Physica, Amsterdam 249–251:961
Schöll E, Landsberg P (1988) Z Phys B72:515
Schöll E, Niedernostheide FJ, Parisi J, Prettl W, Purwins H (1998) Evolution of spontaneous structures in dissipative continuous systems. Springer, Berlin, p 446
Schöll E, Aoki K (1991) Appl Phys Lett 58:1277
Schöll E, Drasdo D (1990) Z Physik B81:183
Schöll E (1992) Handbook on semiconductors, vol 1. North-Holland, Amsterdam
Schöll E, Schwarz G, Wacker A (1998) Physica B Amsterdam 249–251:961
Schöll E (2001) Non-linear spatio-temporal dynamics and chaos in semiconductors. Cambridge University Press, Cambridge
Schöll E (2002) In: Böer KW (ed) Survey of semiconductor physics, Ch 33. Wiley, New York
Schöll E, Hizanidis J, Hövel P, Stegemann G (2007) In: Schimansky-Geier L, Fiedler B, Kurths J, Schöll E (eds) Analysis and control of complex nonlinear processes in physics, chemistry and biology. World Scientific, Singapore, pp 135–183

Schöll E, Majer N, Stegemann G (2008) Phys Stat Sol (c) 5:194
Schöll E (2009) In: Radons G, Rumpf B, Schuster HG (eds) Nonlinear dynamics of nanosystems. Wiley-VCH, Weinheim
Schöll E, Amann A, Rudolf M, Unkelbach J (2002) Physica B 314:113
Schomburg E, et al. (1998) Appl Phys Lett 72:1498
Schomburg E, et al. (1999) Appl Phys Lett 74:2179
Schomburg E, Hofbeck K, Scheuerer R, Haeussler M, Renk KF, Jappsen AK, Amann A, Wacker A, Schöll E, Pavel'ev DG, Koschurinov Y (2002) Phys Rev B 65:155320
Schrepel C, Scherz U, Ulrici W, Thonke K (1997) Mater Sci Forum 258:1075
Schultheis L, Honold A, Kuhl J, Köhler K, Tu CW (1986) Phys Rev B 34:9027
Schultheis L, Honold A, Kuhl J, Köhler K, Tu CW (1986) Superlattices Microstruct 2:441
Schultz O, Glunz SW, Willeke GP (2004) Progr Photovolt: Res. Appl 12:553
Schumacher JO, Altermatt PP, Heiser G, Aberle AG (2000) Solar Energy Mater Solar Cells 65:95
Seeger K (1999) Semiconductor physics, Springer Series on Solid State Science, 6th edn. Springer, Berlin
Schwartz RJ (1982) Solar Cells 6:17
Schwarz G, Schöll E (1996) Phys Stat Sol B 1:351
Schwarz G, Prengel F, Schöll E, Kastrup J, Grahn HT, Hey R (1996) Appl Phys Lett 69:626 and Semicon Sci Technol 11:475
Schwarz G, Schöll E (1997) Acta Tech CSAV 42:669
Schwarz G, Patra M, Prengel F, Schöll E (1998) Superlattices Micro Struct 23:1353
Schwarz G, Lehmann C, Reimann A, Schöll E, Hirschinger J, Prettl W, Novak V (2000) Semicond Sci Technol 15:593, Phys Rev B61:10194
Schwarz G, Lehmann C, Schöll E (2000) Phys Rev B61:10194
Segev M, Collings B, Abraham D (1996) Phys Rev Lett 76:3798
Seitz F (1940) The modern theory of solids. McGraw-Hill, New York, London
Shaw MP, Grubin HL, Solomon P (1979) The gunn hilsum effect. Academic, New York
Shaw MP, Mitin VV, Schöll E, Grubin HL (1992) Physics of instabilities in solid state electron devices. Plenum, New York
Shah J (1996c) Ultrafast spectroscopy of semiconductors and semiconductor nanostructures, Ch 2. Springer, Berlin
Sheng (1990) Scattering and localization of classical waves in random media. World Scientific, Singapore
Shionoya S (1991) Metals Forum 15:132
Shockley W (1950) Electrons and holes in semiconductors. Van Nostrasnt, Princeton, NJ
Silver M, Pautmeier L, Baessler H (1989) Sol State Commun 72:177
Skanavi GI (1958) Phizika Dielectrikov. Fizmatgiz, Moscow
Smith GC (1994) Surface analysis by electron spectroscopy. Plenum, New York
Soares EA, de Carvalho VE, Nascimento VB (1999) Surf Sci 431:74
Spaeth J-M, Meise W, Song KS (1994) J Phys Condensed Mat 6:3999
Spangler J, Finger B, Wimmer C, Eberle W, Prettl W (1994) Semicond Sci/Technol 9:373
Spenke E (1969) Electronic semiconductors. McGraw-Hill, New York
Srivastava GP (2000) Vacuum 57:121

Stegemann G, Balanov AG, Schöll E (2005) Phys Rev E 71:016221
Stegemann G, Balanov AG, Schöll E (2006) Phys Rev E 73:016203
Stegemann G, Schöll E (2007) New J Phys 9:55
Stensgaard I (1992) Rep Prog Phys P55:989
Steuer H, Wacker A, Schöll E (1999) Physica B Amsterdam 272:202
Steuer H et al. (2000) Appl Phys Lett 76:2059
Stier O, Bimberg D (1997) Phys Rev B 55:9740
Stirn K, Böer W, Dussel GA (1973) Phys Rev B 7.4:1433
Stratton R (1969) J Appl Phys 40:4582
Street RA, Mott NF (1975) Phys Rev Lett 35:1293
Stroscio J, Kaiser W (1992) Scanning tunneling microscopy. Academic, New York
Sturge MD (1982) Excitons, Rashba ET, Sturge MD (eds). North-Holland, Amsterdam
Swistacz B (1995) J Electrost 36:175
Symanczyk R, Jäger D, Schöll E (1991) Appl Phys Lett 59:105
Sze SM (1981) Physics of semiconductor devices. Wiley, New York
Sze SM (1985) Semiconductor devices. Wiley, New York
Tanaka K (1998) Jpn J Appl Phys, Part 1 37:1747
Tanaka K, Nakayama SI (1999) Jpn J Appl Phys, Part 1 38:3986
Tani T (1989) Fuji Film, Res Develop 34:37
Tani T (1991) Fuji Film, Res Develop 36:14
Tani T (1995) J Imaging Sci Technol. 39:386
Teitsworth SW, Westervelt RM, Haller EE (1983) Phys Rev Lett 51:825
Tathan MS (1992) Semiconduct Sci Technol 7:B116
Thomas DE, Hopefield JJ, Augustyniak WN (1965) Phys Rev 140:A202
Tolk NH, Traum MM, Tully JC, Madey TE (eds) (1983) Desorption induced by electronic transitions, DIET I. Springer, Berlin
Unkelbach J, Amann A, Just W, Schöll E (2003) Phys Rev E 68:026204
Völkl J (1994) Stress in the cooling crystal. In: Hurle (ed) Handbook of crystal growth. North-Holland, Amsterdam
Valle F, Bogani F (1991) Phys Rev B 43:12049
Van de Walle CO, Neugebauer J (1996) Mater Res Soc Symp Proc 449:861
Vanderbilt D (1990) Phys Rev B41:7893
Varlamov IV, Osipov VV (1970) Sovj Phys Semicond 3:803
Varshneya AK, Seeram AN, Swiler DR (1993) Phys Chem solids 34:179
Vickerman JC (ed) (1997) Surface analysis: the principal techniques. Wiley, Chichester, New York
Vickers AJ, Straw A, Roberts JS (1989) Semicond Sci Technol 4:743
Vinogradov EA, Leskova TA, Ryabov AP (1994) Optics Spectrosc 76:278
Volkov AF, Kogan SM (1967) Sov Phys JETP 25:1095
Volkov AF, Kogan SM (1969) Sov Phys-Usp 11:881
Wacker A, Jauho A-P (1998) Phys Rev Lett 80:369
Wacker A, Schöll E (1994) Z Phys B 93:431
Wacker A, Schwarz G, Prengel F, Schöll E, Kastrup J, Grahn HT (1995) Phys Rev B52:13788
Wacker A, Schöll E (1994) Z Physik B93:431
Wacker A, Schöll E (1995) J Appl Phys 78:7352
Wacker A, Schöll E (1997) J Appl Phys 78:7352

Wacker A (1998) Theory of transport properties of semiconductor nanostructures, Ch 10. Chapman & Hall, London
Wacker A (2002) Phys Rep 357:1
Wagner KW (1926) J Inst Elect Eng 41:1034
Wang CW, Wu CH (1992) Sol State Electron 35:1771
Wang ZL (1996) Reflection electron microscopy and spectroscopy for surface analysis. Cambridge University Press, Cambridge
Ward AJ (1999) Contemporary Phys 40:117
Weber D, Petri W, Woggon U (1997) Phys Rev, B1 55:12 848
Weber MJ, Monchamp J (1973) J Appl Phys 44:5495
Wierschem A, Niedernostheide FJ, Gorbatyuk A, Purwins HG (1995) Scanning 17:106
Weiss S, Mycek MA, Bigot J-Y, Schmitt-Rink S, Chemla DS (1992) Phys Rev Lett 69:2685
Whitehead S (1939) Philos Mag 37:276; Whitehead S (1951) Dielectric breakdown of solids. Clarendon, Oxford, UK
Williams F (1977) J Electrochem Soc 124:C107
Williams F, Karteuser E, Evrard R (1980) Phys Rev B 21:648
Williamson D (1999) Proc Mat Res Soc 557:251
Wolf EL (1975) Sol State Phys 30:712
Wolf T, Ulrici W, Cote D, Clerjaud B, Bimberg D (1993) Mater Sci Forum 143:317
Woodruff DP, Delchar TA (1994) Modern techniques of surface science. Cambridge University Press, Cambridge
Wooten F, Weaire D (1991) In: Ehrenreich H, Turnbull D (eds) Sol State Phys, vol 40. Academic, New York, p 2
Xu H, Amann A, Schöll E, Teitsworth SW (2009) Phys Rev E arXiv:0901.1672
Yamaguchi M, Morifuji M, Kubo H, Taniguchi K, Hamaguchi C, Gmachl C, Gornik E (1994) Sol State Electron 37:839
Yang J, Guha S (1999) Proc Mat Res Soc 557:239
Yarykin N, Cho CR, Zuhr R, Rozgonyi G (1999) Physica B: Condensed Matter 273:485
Yeh P (1988) Optical waves in layered media. Wiley, New York
Yi J, Wallace R, Palmer J, Anderson WA (1994) Solar Energy Mater Solar Cells 33:145
Yu PY, Cardona M (1999) Fundamentals of semiconductors: physics and material properties. Springer, Berlin
Zangwill A (1988) Physics at surfaces. Cambridge University Press, Cambridge
Zhang Y, Yang X, Liu W, Zhang P, Giang D (1994) Appl Phys Lett 65:1148
Zhang Y, Castrup J, Klann R, Ploog K, Grahn HT (1996) Phys Rev Lett 77:3001
Zhao GY, Ishakawa H, Jiang H, Egawa T, Jimbo T, Umeno M (1999) Jpn J Appl Phys, Part 2 38:L993
Zimmermann H, Boyn R, Lehr MU, Schulz H-J, Rudolph P, Kornack J-Th (1994) Semiconductor Sci Technol 9:1598
Zongfu J, Benkum M (1991) Phys Rev 44:11072
Zoorob MF, Chariton MDB, Parker GJ, Baunberg JJ, Netti MC (2000) Nature 404:740
Zou J, Cockayne DJH (1995) Mater Sci Forum 189:279

te Nijenhuis J, van der Wel PJ, van Eck ERH, Giling LJ (1996)J Phys, D: Appl Phys 29:2961
van der Veen JF (1985) Surf Sci Rep 5:199
Zhao J, Wang A, Green MA (1999) Prog Photovolt 7:471
Zhou X (1994) Sol State Electron 37:1888

Index